The Compelling Scientific Evidence for UFOs

Erol A. Faruk, Ph.D

Copyright © 2014 Erol A. Faruk

All rights reserved.

ISBN-10: 150271552X
ISBN-13: 978-1502715524

DEDICATION

To my wife Marina and our daughter Melisa for their encouragement and patience for me to take this project through to its very end.

CONTENTS

    Acknowledgements   vi

    Foreword   vii

    Introduction   xi

1. My initiation into the UFO controversy.   1

2. An overview of the UFO phenomenon   4

    UFOs as a subject of ridicule - UFOs as a subject of secrecy - If UFOs exist, where are the pilots? - If they are here why don't they land and show themselves? - The distances they would need to travel are simply too great - Which physical scientists have been involved in UFO research?

3. The analysis of the Delphos ring soil material and its implications.   41

    Comparison of the Delphos ring with naturally occurring fungal 'fairy rings' - Comparison of Delphos with other physical trace cases

4. Aftermath of ring soil analysis   55

    Phyllis Budinger's analysis of ring soil - Initial approaches to Nature and JBIS Journals - My challenge to SETI chief Seth Shostak - A second attempt to publish in the JBIS - Approaches to publish in the International Journal of Astrobiology and the Journal of Astrobiology - A second attempt to publish in Nature - My concluding communication with Seth Shostak

5. Appendix (original scientific paper submitted for publication)   81

ACKNOWLEDGEMENTS

This book would not have been possible without the participation of the following individuals and organizations to which I am extremely grateful:

Ted Phillips for supplying the original soil material and important information;

Phyllis Budinger for providing additional analysis data on the soil;

Dr. Barrie Bycroft and Dr. Frank Palmer at the University of Nottingham, UK, for providing assistance and enabling me to analyse the soil;

The Mutual UFO Network (MUFON) for allowing the use of the diagrams and photographs from the original case report;

Dr. Mark Rodeghier and Jerome Clark from the Chicago based Center for UFO Studies (CUFOS) for invaluable discussion and help in the preparation of this book.

FOREWORD

Dr. Faruk has taken on an admirable subject; one very worthy to be represented to the intellectual public and the academic world. It focuses on a particular UFO claim of unusual nature, due to the intense scrutiny placed upon it by several major scientific laboratories. The story of the Delphos, Kansas incident represents a nearly heroic effort by these many labs to attempt to dig deep into the UFO mystery, and pry away a little of the layers of questions that all of us have about it.

Dr. Faruk wishes to familiarize readers who have not spent much time themselves digging at this subject with many larger issues involved with UFOlogy, alongside the explication of his own scientific contributions to the case itself. There is no need to refer to the former in this foreword, but, since it may be useful, and is something that I have been mildly involved with myself, I can put the incident itself into its "scientific" and historical context for you.

The case occurred in November of 1971. This was at a unique time in the history of this subject. The University of Colorado "Scientific Study of UFOs" had wrapped up in 1969 amidst a furore occasioned by the odd circumstance that its director, the great physicist, Edward U. Condon, had written a dismissive final conclusion to the report, while twelve of the fourteen major workers on the project were on public record with opposite opinions. Some external scientists [example: Dr. J. Allen Hynek,

Northwestern astronomy professor and advisor to the Air Force, and Dr. James McDonald, University of Arizona atmospheric physicist] had analysed the cases in the report and stated, to their dismay, that even the cases as written in the report didn't support the dismissal. The reasons behind this circumstance are many and complicated, and need quite a bit of reading. If anyone's interested, they could get a copy of "UFOs and Government" (2012) by Robert Powell, Michael Swords, and the UFO History Group, where the entire story is laid out.

The relevance to the current topic is that the Colorado Project ended the USAF's open involvement with UFOs, leaving an investigative gap. Interested persons in the public felt, often, that this was just as well as they had no respect for the USAF typical investigation anyway. Nevertheless, the USAF had money to pursue these things and the public did not. Allen Hynek had experienced enough unknown UFO cases in his 18 years or so of Air Force consultancy that he felt abandoning the subject was an academic crime. Surprisingly, several scientists around the United States felt the same way, and although they didn't wish to focus derisive comments upon themselves, they communicated to Hynek privately that they were interested and would help where they could. These scientists, in the early seventies, nicknamed themselves "The Invisible College". They included physicists, chemists, biologists, NASA grantees, personnel at Oak Ridge, Battelle Memorial Institute, and on and on. It was a potentially formidable group, and one which would turn some of its attention to the Delphos case.

When the Invisible College began communicating with one another, Allen Hynek was wondering how best to proceed with UFO research. His long years with the Air Force had taught him a lot of what NOT to do. He reasoned that it would be the in-depth laboratory science concentrated on special categories of cases, which might not only provide more concrete data [more objective], but also convince the academic community to pay attention. Such high-potential UFO cases were rare but not so rare as to be unavailable. These reports Hynek named as the "Close Encounters of the Second Kind": those cases which had physical impact upon the world. Of this sort of case, it was the so-called "landing trace" case which intuitively seemed the most likely to give lab-bench results.

This sort of reasoning was not complicated and both the USAF and the Colorado Project realized the wisdom of it as well. Neither organization had, however, exploited this type of case. In the Air Force's case it was largely not interested in becoming a science centre for trace case testing, and made little effort to even locate such cases (although in the 1960s plenty existed). In the Colorado situation, organizational incompetence and short timeframe defeated the thought. Hynek had the interest and he had the time. He was almost able to pull off the very difficult organizational problems of a far-flung unofficial pseudo-organization, but what he absolutely lacked was money. All testing would have to be done as "midnight" gratis lab work. Hynek, knowing that money was his Achilles Heel, and that his "collegians" could offer only so much help, decided that a wise target would be only one or two evidence-rich cases per year.

And then along came Delphos.

A more or less local UFO field researcher (and a good one) named Ted Phillips learned of the case reasonably early, did his usual excellent field measurements and sampling, and brought Dr. Hynek in on the "find". Hynek was enthusiastic about the possibilities there, and began marshalling the collegians for a run at the soil samples. Over the year 1972 and into 1973, the universities of Missouri, Kansas, Northwestern, Utah State, Cal-Berkeley, North Dakota, Illinois, and Florida were involved — all anonymously as far as the public was concerned. Persons at NASA, Oak Ridge, and Battelle also weighed in.

As Dr. Faruk will confirm, they found that many things which characterized the soil rendered it different from "untouched" soil nearby, and seemed quite strange. A renowned expert on fungal diseases, Dr. Hubert Lechevalier of Rutgers, was particularly valuable, as he could claim that the ring, though containing some fungal elements as almost all soils do, was NOT the product of such an agency.

But the mystery, even of the exact physical constitution of the ring soil, remained elusive. Some years passed, and Dr. Faruk, as he will describe to you, became involved in the mystery. Soil samples were still in good shape and made available to him. And there his part in this story began. Delphos is a real scientific saga. It's a serious saga aimed at a mystery — a mystery having utterly legitimate science components. Dr. Hynek and the Invisible Collegians approached it as true scientists, open and ready for the quest. Dr. Faruk has done the same.

Enjoy his book.

Michael D. Swords,

Professor Emeritus of Environmental Studies and Natural Sciences,

Western Michigan University.

INTRODUCTION

This book has been written for the purpose of bringing to the attention of the general public a UFO event that I believe to be of great importance. It is often claimed by sceptics and debunkers that there is no reliable evidence for the existence of UFOs. This volume takes the opposite standpoint, and presents the data obtained from a chemical analysis that I personally undertook on soil samples derived from a farm in Delphos, Kansas on which a close proximity UFO was observed hovering just above the ground. Although some of this information was originally published in 1989, the journal it appeared in (Journal of UFO Studies) had a relatively small readership and therefore has not been generally accessible to the public. Corroborative data from the ring soil has since been independently obtained and published elsewhere by another analyst, Phyllis Budinger in the USA.

Attempts have been made more recently to publish an updated version of my findings in a fully peer reviewed science journal such as Nature, the Journal of the British Interplanetary Science (JBIS), The International Journal of Astrobiology, or Astrobiology journal in order for it to receive a much wider scientific readership. Despite the title of the updated paper: *The search for signs of extraterrestrial intelligence on earth: strong chemical and physical evidence for the existence of an unconventional luminescent aircraft (commonly called a UFO) observed by multiple witnesses at a farm in Delphos, Kansas, USA* none of these journals accepted it for publication, on the basis that the subject matter of the paper was deemed 'inappropriate'. This was not altogether surprising, since the subject of UFOs is still unfortunately considered to be something of a taboo for scientists to discuss in general. This book will also address why this taboo exists and explain why it is greatly misplaced.

The observation that scientists should currently be making enormous efforts through SETI and the like to discover if life exists elsewhere in the universe but adamantly - and ludicrously - refuse to investigate UFO reports seriously will also be considered.

Part 1. My initiation into the UFO controversy

Perhaps the best place to start is how I became involved with the subject of UFOs in the first place. I was born in London in July 1951 from Turkish parents who emigrated from Cyprus in 1948. My father was an accomplished tailor who found employment as a bespoke tailor fairly quickly and was therefore able to feed and house the family in an area of North London called Kilburn. I had a fairly normal upbringing from caring parents, the only caveat being that I spoke little English before entering the local Kingsgate Primary School at the age of 5 or thereabouts. However, I quickly learned it at school - as one does at that age - and thereby acquired a bilingual ability that has served me well throughout my life.

When I was about seven years old I asked my mother to buy me a toy microscope because of my curiosity to see small things magnified. One of the enclosed specimen samples was that of a tiny insect called 'Thrips' (commonly known as a Thunder Fly) which I was astonished to see become a repulsive monster once viewed under the microscope's eyepiece. As I grew older my fascination with optical equipment then encompassed telescopes in order to view the skies. I had become awed by the immensity of the universe and was an avid visitor to our local library searching for Patrick Moore's many popular books on amateur astronomy. I once managed to follow his instructions to build a refracting telescope using cardboard tubes and a three inch chemist's watch glass which fortuitously had the desired convex shape to produce a sharp focus at approximately three feet or so. My first views of the moon's craters were obtained using this telescope. I eventually persuaded my father to purchase a larger equatorially mounted Newtonian reflecting telescope to boost my viewing of the heavens, and I retain an interest in the hobby even today. It was while I was perusing through books in the astronomy section of the local library that I chanced upon one on UFOs. This was written by Desmond Leslie called "Flying Saucers have Landed", which he co-authored with the contactee George Adamski and published in 1953. I was now in my teens and was able even then to view Adamski's outlandish claims of meeting with Venusians with great suspicion. It was the contents of Desmond Leslie's section of the book which impressed me however, and it was very obvious to me even then that the observations of flying objects with very similar appearances and characteristics being reported from across the world by people of widely differing cultures couldn't all be manifestations of mass hallucination. I became eager to study the UFO phenomenon from that period onwards.

I remember briefly meeting Patrick Moore at a British Astronomical Association meeting once (I had become a junior member, albeit for a short period). Although I don't remember what I spoke to him about, I did then use the opportunity to write to him and remind him that we met while enquiring what his views were on UFOs. He was kind enough to reply with a postcard but was rather dismissive of UFO reports and didn't take them at all seriously, suggesting that they had about the same credibility as the Flat Earth Society. It was a shock to discover someone whom I looked up to betraying an irrationally negative attitude towards UFOs.

My early interest in science meant that I left school with four Advanced Level GCSEs in Chemistry, Physics, Pure Maths and Applied Maths, with a top grade in the former subject. My enthusiasm for chemistry inevitably led to my doing experiments at home such as preparing a spool of crude nylon, as well as dabbling with homemade fireworks which didn't exactly enamour me to our next door neighbour!

I went on to read chemistry at Queen Mary College, London University and after receiving my first degree stayed on to take a

Ph.D. in the organic synthesis of air-sensitive carotenoid pigments. This was followed up with research posts at Oxford and Nottingham Universities for a total of just over three years, before I eventually found employment as a development chemist at Beecham Pharmaceuticals in Harlow, Essex (which later enlarged to become GlaxoSmithKline Pharmaceuticals).

During these years my interest in UFOs never waned, and I tried to keep up-to-date with events across the world by subscribing to the premier journal of that period *Flying Saucer Review* (FSR), published in the UK. It was during a period of greater influx of UFO reports that the editor of the FSR decided to publish a sister journal called *FSR Case Histories*, which I naturally also subscribed to. It was in the Supplement 9 issue (February 1972) of this particular journal that I first came across the so-called Delphos Case from Kansas in the USA that involved a brightly glowing UFO hovering beneath a tree before departing to leave a glowing ring in the ground. The latter was photographed and found to exhibit very peculiar water repellent properties that greatly intrigued me. I waited to see whether further information on these soil peculiarities would be published but nothing significant emerged.

A few years later I was undertaking postdoctoral research at Nottingham University when I decided to contact Dr. Mark Rodeghier at the Center for UFO Studies (CUFOS) in Chicago through an intermediary to see whether I could get hold of some of the affected soil material. He, in turn, contacted the principal investigator of the case Ted Phillips who duly sent me several grams of the material held within airtight opaque containers that would normally hold photographic film. I recall Rodeghier mentioning in one of his communications that both Phillips and Dr. J. Allen Hynek, the founder of CUFOS, were very keen for me to analyze the soil.

I proceeded to examine the ring soil to try and uncover what was causing the water repellent behaviour. I half expected to find something that would be easily explained in down-to-earth terms. I was therefore pleasantly surprised and excited to discover an air-sensitive compound having rather special qualities that eventually led me to conclude that its presence actually supported the alleged UFO sighting. This aspect of the story will be discussed more fully later, while the actual details of my chemical analysis and its conclusions are presented in the scientific paper that forms the subject of this book (see the Appendix). But first - for the benefit of those readers having perhaps only limited knowledge - I would like to provide some background information and personal opinions on this controversial subject.

Part 2. An Overview of the UFO Phenomenon.

This will necessarily be brief since there are numerous books available on UFOs that cover its scope and history very well. Although there have been many instances of unusual aerial phenomena being recorded throughout the millennia, the modern era is considered to have begun during World War II, when allied bomber pilots reported seeing glowing balls of light flying alongside them, apparently monitoring their activities. The lights were nicknamed 'foo-fighters' by the pilots. Two books published relatively recently have delved into this early phase of the UFO phenomenon which should be invaluable to those interested: *UFOs in Wartime: What They Didn't want you to Know* by Mack Maloney (2011) and *Strange Company: Military Encounters with UFOs in World War II* by Keith Chester (2007).

Following the war sightings began to be reported of flying metallic discs across Europe and the United States which led to the Roswell 'saucer crash' incident in 1947 and the mass sightings over Washington D.C. in 1952. The latter, which formed part of a larger international wave, generated great public interest and even led to a Presidential decree to 'shoot them down'! More information on this wave can be found at this link:

http://www.nicap.org/waves/1952fullrep.htm

From that early period the phenomenon has developed into a global affair with hundreds of thousands of reports being accrued of unfamiliar 'structured craft' that form the core of the phenomenon. It must be emphasized that this core isn't about vague 'lights in the sky' that could easily be misinterpretations of astronomical objects or terrestrial aircraft, but are instead reports of sizeable 'craft' with wholly unconventional designs and extraordinary performances that suggest that they do not originate from our planet! Many of these reports are anecdotal in nature which forms the basis of the scientific skepticism that is still prevalent today, since without confirmatory photographic or radar based data they can easily be called into question. Clearly, a UFO sighting by just one person will have severe credibility issues. It's the many cases where multiple witnesses are involved - sometimes hundreds or even thousands of people - reporting the same object or objects seen from different vantage points that necessarily will have far greater credence attached to them. There have been several instances where such mass sightings have taken place, the 1952 Washington D.C. event mentioned above being an early example. I will give two further examples.

The first is the famous Belgian wave of sightings of mainly large triangular shaped craft with lights, which lasted from 29 November 1989 to April 1990. This peaked on the night of 30/31 March 1990 when unknown objects were tracked on radar, chased by two Belgian Air Force F15 jets, and sighted by an estimated 13,500 people on the ground, 2,600 of whom filed written statements describing in detail what they had seen. Witnesses included military and police officers, pilots, scientists and engineers. Following the incident the Belgian air force released a report detailing those events while the Belgian government cooperated with civilian UFO investigators, an action without precedent in the history of government involvement in this field. The Head of

Operations of the Belgian Air Staff at the time was a colonel Wilfried De Brouwer who was directly involved in collating the strange events and became convinced that something extraordinary was taking place.

In 2010 an excellent and very informative book on UFOs was published authored by Leslie Kean, an investigative journalist, entitled *UFOs: Generals, Pilots and Government Officials go on the Record*. One of the chapters in this book was written by De Brouwer himself (now a retired Major General) in which he describes in some detail a number of the events that took place during that period, including the observation by two policemen patrolling the road of a large triangular craft hovering over a field with three beaming spotlights at its corners and a red light at the centre. The object moved around purposefully and was eventually joined by a *second* similar object that appeared from behind woods later on. This was only one of many police sightings of strange aerial objects at the time, and the interested reader would do well to read De Brouwer's full account in Kean's book, a link to which is provided here:

http://www.ufosontherecord.com/

My second example is the equally famous mass sightings of a huge and silent boomerang object with 5 - 7 lights on its 'wings' reported over the skies of Phoenix, Arizona in March 13, 1997. These started at around 8pm at the north end of the state and proceeded southwards towards the city itself and involved hundreds, possibly thousands of witnesses. Many reports came in between 8.00-10.00pm as people went outside of their homes to observe the Hale-Bopp comet that was also observable at the time. The police were inundated with telephone calls describing the craft and some officers ended up going out to view it for themselves. Even the Governor of the state observed the object and was subsequently interviewed by a film maker, James Fox, who went on to create a documentary that incorporated several key witness testimonies of the giant craft. The documentary in question is entitled *I Know What I Saw* and can be sourced through Amazon. Segments can sometimes be found on YouTube on the Internet. In my opinion it is one of the best documentaries on UFOs ever produced, and I would urge the reader to watch it for themselves to gauge the credibility and sincerity of the many witnesses interviewed.

Another DVD-based documentary on the sightings which is well worth watching is *The Phoenix Lights: Beyond Top Secret* produced by Dr. Lynne D. Kitei. She was a witness herself of a similar object *before* March 13th and even took photographs of a V-shaped object. This documentary, which has won numerous film festival awards, includes many other witness testimonies of the same craft observed on March 13 and has bonus features such as positive commentary on UFOs from Dr. Edgar Mitchell, one of the astronauts that took part in the moon landings. Although the main sightings of the boomerang craft over Phoenix were reported between 8.00 and 10.00pm, the U.S. Air Force did its best to create confusion over the origin of the reports by dropping flares just after 10pm which naturally the sceptics pounced upon to propose as an 'explanation' for the bulk of the sightings. It is very evident from watching the documentaries that the interviewed witnesses themselves resolutely did not agree with the flares suggestion! Video footage of the actual flares dropped by the planes which show them slowly descending and then being obscured by distant hill tops can be sourced on the internet. It must be emphasized that these video clips have nothing to do with the earlier sightings of the large boomerang object. The motive of the U.S. Air Force for dropping the flares is likely to be based on their familiar 'smokescreen' policy of trying to minimize public interest in the phenomenon which will be discussed more fully below.

Multiple witness sightings are not the only ones that have much greater credibility – many others are coupled with simultaneous radar tracking of unusual objects, and there are by now over 5000 cases where landing marks or traces have allegedly been left behind by a UFO, the title Delphos case being an exceptional example.

UFOs as a subject of ridicule

If someone mentions UFOs in any public gathering the response that is likely to be elicited is one of embarrassed giggling or even disdain, as if they were referring to a belief in fairies or the Flat Earth Society as Patrick Moore had opined. I believe there are two main reasons why this response is evoked.

Firstly, the early history of the involvement of the U.S. Air Force into investigating UFOs led to it being dragged into an area of high controversy that it wished to avoid, and consequently it was quick to pour cold water on those sighting reports that were of notable public interest. What better – and convenient - way of explaining away a report was there than by casting aspersions on the credibility of the reporter? Since people were seeing things that didn't formally exist, it was easy to propose they were actually suffering from some form of hallucination. Once the media latched onto this idea there was no going back, and journalists generally delighted in treating such reports with scorn and ridicule, to the extent that the reporting of many newsworthy sightings were subsequently bitterly regretted by their observers as evidenced from recorded TV and news coverage of cases from the 1950s and 60s. This kind of media derision is unfortunately still prevalent today, and appears to be the preferred treatment when sightings of UFOs are mentioned in newspapers.

Secondly, the subject of UFOs – by their implied origin – leads to an uncomfortable worldview that elicits a strong 'fear of the unknown' element that many people would rather not face up to. The notion of likely alien hostility has been drummed into the human psyche for the last 100 years or so, starting with H.G. Wells' novel *War of the Worlds* and continued with countless science fiction films portraying malicious visitors from outer space whose main preoccupation was the extermination of human life. A subjective fear of alien visitation is therefore quite understandable, and rather than question the validity of UFO sightings it is far easier for many just to sweep the idea of such under the carpet and wilfully ignore it. It is very possible that the irrational hostility displayed towards UFOs by many scientists (Patrick Moore would be an example) come under this category. Outwardly it is certainly puzzling that instead of eagerly studying UFO reports to fulfil their naturally enhanced curiosities, scientists generally take gleeful pleasure in summarily dismissing them out of hand without even bothering to look at the evidence – a thoroughly unscientific approach if ever there was one! Admittedly the evidence that science normally requires to undertake an investigation would be of a type that could be readily analyzed in the laboratory – and clearly UFOs generally do not fall into that category. But that should just mean that there is a difficult problem to tackle. A good analogy here would be the refusal by eighteenth century scientists to accept that stones could fall from the sky. A German scientist, Ernst Chladni, was the first to propose the idea in a booklet published in 1794, and this led to much mockery by his peers. It took another ten years and work by a French scientist, Jean-Baptiste Biot who examined an avalanche of meteorite falls in a localized area of France in April, 1803 that finally led to the acceptance of the phenomenon. The moral of this story must be that, when the need arises, scientists will have to be proactive and go out and investigate for themselves a phenomenon that cannot easily be brought into the laboratory.

UFOs as a subject of secrecy

The early intervention of the US Air Force to disclaim the Roswell, New Mexico 'crash' incident allowed the air force to conveniently ignore the matter. The Roswell event is certainly controversial, and debates still occur among seasoned ufologists about what actually took place there in June, 1947. The fact remains that an aerial object of some kind had crashed leaving a field strewn with debris of an unusual type. This material was initially discovered by a ranch foreman Mack Brazel who reported his finding to the local sheriff. The latter then contacted the Roswell Army Air Field (RAAF) which sent out two officers Major Jesse Marcel (the base intelligence officer) and Captain Sheridan Cavitt (the head of counterintelligence) to recover and examine the material. On July 8 the RAAF information officer issued a statement explaining that their personnel had recovered a 'flying disk' which had crashed on a ranch near Roswell which the local press then reported (see figure below). Shortly after, however, an air force officer, General Roger Ramey appeared on the scene and issued another statement dismissing the event as having been caused by nothing more sinister than a weather balloon! A photograph was then taken of Major Marcel handling material from such a weather balloon and presented to the media as conclusive evidence for the balloon explanation. The event was then quickly forgotten as far as the public was concerned. It was much later, in the late 1970s, that certain investigators – notably Stanton Friedman (a nuclear physicist) and Bill Moore – rediscovered the story and proceeded to uncover valuable testimony from many other people concerning the event.

Numerous books and documentaries have since been produced and the name Roswell has acquired prominence as *the* leading UFO cover-up by the U.S. authorities. Although much has been written on the matter, two relatively recent books can be recommended for those wishing to garner a fresh insight into the event. The first of these is *The Roswell Legacy – The Untold Story of the First Military Officer at the 1947 Crash Site*. This is written by Dr. Jesse Marcel Jr. (the son of Major Marcel), and his wife Linda Marcel.

Local newspaper headline of Roswell 'crash'

The book, with a foreword by Stanton Friedman, is a detailed account of Dr. Marcel's own knowledge and memories of the event and describes, for example, how Major Marcel had arrived home in an excited state and showed his then eleven year old son and wife some of the very unusual material that he had recovered from the ranch. The book also explains why Major Marcel was manoeuvered into appearing in a sham media event to promote the weather balloon idea. Dr. Marcel is adamant that the Roswell crash is a genuinely anomalous event and makes clear that his book is an attempt to set the record straight as well as keep the promise he made to his father to enable the latter's own story to be told. The second book is *Witness to Roswell, Revised and Expanded Edition: Unmasking the Government's Biggest Cover-Up* by Thomas Carey and Donald Schmitt. This has an extensively researched account of the many people that were directly affected by the Roswell event, with the implication that some of them were silenced through fear in order not to discuss the matter. The book has a foreword by the former astronaut and Apollo 14 moon walker Dr. Edgar Mitchell who actually grew up in the Roswell area. After becoming celebrated as an astronaut he returned to Roswell and met some of the people ('the old-timers') who had knowledge of the crash, and he himself clearly believes it involved an extra-terrestrial vehicle and subscribes to the view that there is major U.S. government cover-up of UFOs. There is a web link to an interview he gave on the UK Kerrang Radio Station on July 23rd 2008 in which he openly expressed these opinions - to the obvious astonishment of the interviewer:

www.mufonnd.webs.com/apps/videos/show/7439276

Although it publicly disowned the Roswell event it is evident from other documentary evidence of that period that the U.S. Air Force was keenly interested in the UFO phenomenon. On September 23, 1947, General Nathan Twining (who was eventually promoted to Chairman of the Joint Chiefs of Staff) sent a memo to Brigadier General Shulgen offering the opinion of Air Materiel Command on UFOs. A transcript of part of this memo is reproduced below (the original can be found on-line in PDF form, and was obtained through the Freedom of Information Act - FOIA). (Refs: www.afterdisclosure.com/2010/09/twining-memo-1.html; www.roswellfiles.com/FOIA/twining.html):

SUBJECT: AMC Opinion Concerning "Flying Discs"

TO: Commanding General

Army Air Force

Washington 25, D.C.

ATTENTION: Brig. General George Schulgen

AC/AS-2

As requested by AC/AS-2 there is presented below the considered opinion of this command concerning the so-called "Flying Discs." This opinion is based on interrogation report data furnished by AC/AS-2 and preliminary studies by personnel of T-2 and Aircraft Laboratory, Engineering Division T-3. This opinion was arrived at in a conference between personnel from the Air Institute of Technology, Intelligence T-2, Office, Chief of Engineering Division, and the Aircraft, Power Plant and Propeller Laboratories of Engineering Division T-3.

2. It is the opinion that:

a) The phenomenon is something real and not visionary or fictitious.

b) There are objects probably approximating the shape of a disc, of such appreciable size as to appear to be as large as man-made aircraft.

c. There is a possibility that some of the incidents may be caused by natural phenomena, such as meteors.

d. The reported operating characteristics such as extreme rates of climb, manoeuvrability (particularly in roll), and motion which must be considered <u>evasive</u> when sighted or contacted by friendly aircraft and radar, lend belief to the possibility that some of the objects are controlled either manually, automatically or remotely.

e. The apparent common description is as follows:

1) Metallic or light reflecting surface.

2) Absence of trail, except in a few instances where the object apparently was operating under high performance conditions.

(3) Circular or elliptical in shape, flat on bottom and domed on top.

(4) Several reports of well kept formation flights varying from three to nine objects.
(5) Normally no associated sound, except in three instances a substantial rumbling roar was noted.
(6) Level flight speeds normally above 300 knots are estimate

---

The reader should particularly note opinion 2(a) in the above that "The phenomenon is something real and not visionary or fictitious."

The impact of this important memo eventually led to a formal Air Force investigation of UFOs called Project Sign that took place during 1947-1949. This in turn led to the issuing of a document written by Sign's staff in late 1948 called the "Estimate of the Situation" which concluded that UFOs were most likely extraterrestrial in origin. This was then forwarded to the office of

General Charles Cabell, the chief of Air Force intelligence. The document eventually passed into the hands of General Hoyt Vandenberg, Chief-of-Staff of the U.S. Air Force, who requested a revised estimate due to lack of supporting physical evidence for UFOs. However, Sign personnel refused to abandon the interplanetary hypothesis, leading many to be reassigned, and in 1949 Project Sign was renamed Project Grudge and then staffed by people who were decidedly antagonistic towards the extraterrestrial hypothesis. It issued its only formal report on UFOs in August 1949, the conclusions of which are shown below. Project Grudge was then disbanded.

There is no evidence that objects reported upon are the result of an advanced scientific foreign development; and, therefore they constitute no direct threat to the national security. In view of this, it is recommended that the investigation and study of reports of unidentified flying objects be reduced in scope. Headquarters AMC [Air Materials Command] will continue to investigate reports in which realistic technical applications are clearly indicated.

NOTE: It is apparent that further study along present lines would only confirm the findings presented herein. It is further recommended that pertinent collection directives be revised to reflect the contemplated change in policy.

All evidence and analyses indicate that reports of unidentified flying objects are the result of:

1. Misinterpretation of various conventional objects.

2. A mild form of mass-hysteria and war nerves.

3. Individuals who fabricate such reports to perpetrate a hoax or to seek publicity.

4. Psychopathological persons.

When General Cabell discovered that Project Grudge had summarily dismissed UFO reports he helped to reinstate Grudge under the new name of Project Blue Book in 1952, which under the initial direction of Air Force Captain Edward Ruppelt developed into a more open minded and productive organization. This is when the term UFO was originally introduced to signify a neutral stance on the nature of the flying objects. Over the next 18 years a total of 12,618 UFO reports were collected and analyzed, with Blue Book concluding that the majority of the sightings were misidentifications of natural phenomena or conventional aircraft. However, a small percentage remained unexplained, even after careful analysis.

In 1968 the only major scientific study on UFOs ever conducted led to the infamous Condon Report being published, which concluded that there was nothing anomalous about UFOs. Project Blue Book was therefore ordered to terminate its activities in December 1969, and since then there has not been any official U.S. government body set up to investigate UFOs that the public is aware of. The Condon Report was itself seriously flawed, since its negative conclusion was, in essence, the personal opinion of its head, the physicist Edward U. Condon, who was overtly dismissive of the extra-terrestrial hypothesis for UFOs. In fact, just a year before publication, Condon openly stated in a lecture that he thought the government should not be studying UFOs because he thought they were nonsense, adding, "but I'm not supposed to reach that conclusion for another year". And this despite the fact that approximately 30% of the reports described in the Condon Report remained unexplained even after analysis by its scientists!

Although the above doesn't provide any hint of a more serious – and possibly covert – study into UFOs by the U.S. authorities, consider the following statement made by retired U.S. Vice Admiral Roscoe Hillenkoetter which was published in the highly regarded New York Times of February 28th, 1960:

"Behind the scenes, high ranking Air Force officers are soberly concerned about UFOs. But, through official secrecy and ridicule, many citizens are led to believe the unknown flying objects are

nonsense... to hide the facts, the Air Force has silenced its personnel."

Hillenkoetter had a distinguished naval career during World War II and then became the first director of the Central Intelligence Agency that was created by the National Security Act of 1947. He was with the CIA until 1950 after which he returned to naval duties to command a cruiser division in the Korean War for which he was promoted to Vice Admiral before retiring in 1957. And what does Hillenkoetter do after retiring? He joins the governing board of the National Investigations Committee On Aerial Phenomena (NICAP) – a civilian organization set up the year before to investigate UFO sightings! Here is someone having an impeccable military background who was then appointed by President Truman to become the first director of the CIA and who then ultimately gets involved in UFO research! It doesn't need much to contemplate that a former director of the CIA is likely to have known a great deal about covert U.S. government investigation into UFOs, leading to his surprising statement to the media in 1960.

Now consider the fact that in 1999 a major report was published by a high level French group called COMETA that had studied the possible implications of UFOs for the defence of that country. The report concluded that a small proportion of the cases they investigated were inexplicable and that the best hypothesis to explain them was extraterrestrial visitation. The people involved in conducting the study were high-ranking military men and officials, some having held command posts in the armed forces and aerospace industry, and were mostly ex-members of the Institute of Advanced Studies for National Defence, a French military think-tank. Although the report was not authorized by the French government its publication nevertheless drew considerable media attention across the world and led directly to the interest in UFOs by the investigative reporter Leslie Kean, whose book was referred to earlier. In her publication there is a very interesting interview with the president of COMETA, Air Force General Denis Letty. Significantly, COMETA also hinted at a massive cover-up of UFO evidence by the U.S. authorities. An English translation of the

COMETA report is accessible on-line while an overview can be read here:

http://www.cufos.org/cometa.html.

In 2000 a book entitled *You Can't Tell the People* was published which told the story of the Rendlesham Forest UFO incident that occurred in December 1980. This is perhaps the UK's most famous case involving military personnel at a U.S. Air Force base and repeat sightings of UFOs with beams of light directed towards the ground and at nuclear weapons facilities. The book was authored by yet another investigative journalist, Georgina Bruni. While researching the case she questioned Baroness Margaret Thatcher on the incident because the latter was Prime Minister during the time it occurred. The Baroness's riposte is reported to have been "UFOs! You must get your facts right and you can't tell the people" from which the title of the book was derived. Bruni later became associated with Admiral of the Fleet Lord Hill-Norton in her quest to get to the bottom of the UFO incident, with Hill-Norton succeeding in asking questions in the House of Lords regarding the case as well as on possible secrecy issues over UFOs.

The reader may be interested to learn that there is also video testimony from several ex-military personnel in the U.S. that even implicate UFOs in the partial disabling of nuclear missile silos by apparently unknown means! A leading researcher in this area is Robert Hastings who has a dedicated website on the matter (ufohastings.com) to which the reader is referred. It is yet another piece of the jig saw which suggests that UFOs are being taken very seriously by the U.S. military. Hastings has written a number of articles on the missile issue, two of which can be accessed here:

http://www.cufos.org/missiles.pdf

http://www.cufos.org/hastings.pdf

If UFOs exist, where are the pilots?

If there's one thing that's guaranteed to elicit more hostility and scepticism towards UFOs it's the reports of aliens that are sometimes featured with them. Logically, of course, there shouldn't be any surprise at the association. I first came to realize the sheer extent of such reports by reading a special edition of the FSR called *The Humanoids* that had page after page of incidents gleaned from newspaper clippings from around the world. Typically these consisted of sightings of landed saucers from which the occupants emerged to reconnoitre the area or apparently take part in some sort of repair or servicing activity on the landed craft. These early reports mainly originated from Europe and South America, and often characterized the aliens as being humanoid in appearance but diminutive in stature (i.e. 3-4ft tall) and wearing some sort of close fitting overalls.

While the interaction of these occupants with humans was initially low key, reports later emerged of alleged abductions wherein humans were taken on board UFOs against their will for physical and/or medical examination. The first such case to be widely publicized was the Betty and Barney Hill abduction that occurred on the night of September 19, 1961 in New Hampshire, USA. The couple were travelling back to their home by car from a vacation when they saw a bright star like object in the sky that followed and then descended towards them, getting larger as it did so. Eventually the craft hovered in front of their car filling the windscreen with its size and causing them to come to a halt in the middle of the highway. Barney described the craft as huge and looking like a giant pancake. Using a pair of binoculars he could see humanoid figures looking down at them through windows of the craft at which point he panicked and suggested to Betty that they should get away from the object as fast as possible. As they drove away they experienced a tingling sensation followed by a missing time episode in which they unexpectedly found themselves 35 miles further down the road. Not knowing what had happened the couple sought help and were eventually referred to a medical practitioner Dr. Benjamin Simon who hypnotized Betty and Barney separately to discover their missing experiences. Under hypnosis they each described - and with considerable emotion - having been taken on board the craft and medically examined, with Barney believing that a sperm sample had been taken. They were then returned back to their car and allowed to proceed with their journey but only after their memories of the experience had been erased. This is a well known case which has been extensively written about and even depicted in a number of television mini film series. A book that can be recommended for its sheer detail is *Captured! The Betty and Barney Hill UFO Experience: The True Story of the World's First Documented Alien Abduction* authored by Stanton Friedman and Kathleen Marden, the latter being Betty Hill's niece. The event also represents an archetypal experience that has similarly been reported by a great number of other alleged abductees ever since. Examples that can be considered to have merit because of the sheer numbers of people involved include the Travis Walton case (a forest worker incapacitated by a UFO in front of five colleagues before being abducted for several days) and the Allagash abduction (a group of four campers taken and examined onboard a UFO before being released with their memories erased of the event).

Whether or not these experiences can be classed as physically 'real' events is unknown at this time. But they are certainly very much a part of the wider UFO mystery. An early pioneering investigator of these abduction cases was the late Budd Hopkins, a professional artist, who started investigating UFO sightings after experiencing his own in 1964. He learnt of encounters that led to missing time episodes similar to that described by Betty and Barney Hill. Realizing that these could implicate events of deeper significance he initially employed professional hypnotherapists to try and uncover those memories from the claimants that were missing. These invariably gave rise to harrowing accounts of abduction on board a UFO leading to some form of medical examination. In 1981 Hopkins published his ground breaking volume *Missing Time* which described a number of different abductions having strikingly similar characteristics, and in 1989 set up the Intruders Foundation, a non-profit investigational organization that also provided moral support for the experiencers. There is a web link to an interview with Hopkins that reveals his thoughts and insights into the abduction phenomenon:

http://www.pbs.org/wgbh/nova/aliens/buddhopkins.html

Although Hopkins initially employed professional hypnotherapists to conduct the hypnosis sessions he eventually became astute in performing these himself, which however led to criticism that he might subconsciously be instilling false memories into his subjects through the power of suggestion. To counter this he went to the trouble of inviting a number of professional psychiatrists to oversee the sessions in order to demonstrate that he wasn't leading the abductees in any way through his questioning. One of the psychiatrists that took part was Professor John E. Mack from the Harvard School of Psychiatry. Initially very sceptical, he became so intrigued by observing that unrelated abductee experiences followed a certain consistency that he himself became heavily embroiled in the investigation of the phenomenon, much to the chagrin of his Harvard peers. He wrote two bestselling books, *Abduction: Human Encounters with Aliens*, and *Passport to the Cosmos*, details and reviews of which can be found here:

http://experiencers.com/books-by-john-e-mack/#abduction

Sadly, Professor Mack was accidentally killed by a drunk driver while in London to present a lecture in September, 2004. Much exists on the web regarding his investigation into abductions, including this interview with him:

http://www.intuition.org/txt/mack1.htm

A link for an overview of his abduction research can be accessed here:

http://makemagicproductions.com/johnmack/

Another leading researcher in the field is Dr. David Jacobs, a recently retired Associate Professor of History who has also used hypnosis extensively on alleged abductees to try and establish the veracity and meaning of their experiences. Dr. Jacobs originally came into prominence with the publication of his first book *The UFO Controversy in America* in 1975, and in 2000 he edited *UFOs and Abductions: Challenging the Borders of Knowledge* which was a scholarly review by several scientists and investigators of the phenomenon at that time. In order to appreciate Jacobs' conviction of the importance of studying the abduction phenomenon here is a review he wrote of a book written by Dr. Susan Clancy entitled *Abducted: How People Come to Believe They Were Kidnapped by Aliens*:

http://www.scientificexploration.org/journal/reviews/reviews_20_2_jacobs.pdf

One intriguing feature of these abduction narratives is the prevalence of a uniformly lit room in which the abductees often find themselves in once on board the UFO. The description that is often used to portray the experience is that this illumination 'comes from everywhere'- i.e. suggesting that it comes from the walls themselves. This is a most interesting observation and by itself lends a degree of modest corroboration to the abduction experience that was first highlighted in the pages of FSR. Curiously, it may also have relevance to the findings from the Delphos ring soil analysis which will be discussed shortly.

If they are here, why don't they land and show themselves?

This is a valid question, and one that I have often pondered about. Since the phenomenon tends to be elusive and mainly nocturnal as far as encounters with humans are concerned, it is apparent that a level of secrecy is their preferred modus operandi. If UFOs are extra-terrestrial in origin then it's likely that any overt contact in the form of a very public landing in a crowded city would lead to a degree of panic by the latter's citizens which could conceivably spread much further. No one can possibly predict what would happen in such circumstances, and it may well be that the drip, drip encroachment by UFOs into our society over many years is their way of lessening the shock of eventual full contact – i.e. imparting necessary conditioning for human kind.

Of course, it may *not* be their intention to develop such open contact in the first place! They may just wish to study human activities from afar, as we do using cameras to record wildlife in their natural habitats. If we are to consider seriously the many alleged abduction experiences reported over the years there appears to be an interest in human reproductive systems in view of the harvesting of human sperm and eggs that is frequently claimed by abductees. What this might mean is impossible to determine at present, and further speculation is best avoided until more meaningful evidence becomes available.

The distances they would need to travel are simply too great.

I am naturally aware of the huge distances that need to be traversed between stars and their respective planetary systems if UFOs were indeed alien in origin. The notion that such journeys are all but impossible is *the* main factor for the strong scepticism displayed by scientists regarding the reality of the phenomenon. But we also previously 'knew' that stones couldn't fall from the sky – until confirmatory evidence for such was eventually discovered in 1803! Our scientific heritage is approximately 400 years old. What are we likely to achieve in another 1000 years, or indeed 10,000 years?

In recent years the possibility of there being multiple universes coexisting with our own has been seriously proposed by leading scientists. There has also been much discussion about the nature of both Dark Matter and Dark Energy, the pervading presence of which have been indirectly established through astronomical observations. Clearly, our knowledge of the universe and its underlying nature are currently incomplete, and therefore any doubts regarding the possibility of interstellar travel by an alien civilization that may be thousands of years ahead of us are likely to be premature. A clue to this can be gauged from the following report. On November 7, 1986 the famous case of the Japan Air Lines flight 1628 encounter with a huge UFO occurred near Anchorage, Alaska at around 5pm. The object resembled two soup plates attached rim to rim with lights running around it and was estimated to be as large as an aircraft carrier. The plane was a cargo carrying Boeing 747, which only had the pilot and two crew members on board. They all observed the object - visually as well as on radar. Over a half hour period the UFO performed remarkable manoeuvres, such as at one point appearing about eight miles in front of the plane, and then suddenly repositioning itself seven miles *behind* the plane in a matter of a few seconds, with radar confirmation of the proximity change. The captain of the plane described the displayed technology as "unthinkable" because of the apparent control of the object over both inertia and gravity. This case is described fully in Kean's book, and I refer to it here because the abruptness of the change in the UFO's position indicates that, in some instances, they do not merely 'fly' to traverse space.

Which physical scientists have been involved in UFO research?

Although the scientific community has so far largely ignored the UFO phenomenon it would be instructive to look closer and reveal just how readily those scientists who do take the trouble to become acquainted with it end up becoming vocal advocates for its active investigation. Mention has already been made on how Professor John E. Mack, a Harvard psychiatrist, had gone from being a total sceptic on abductions to a leading investigator of the same. Such about-turns can also be cited within the physical sciences, a very notable example being Dr. J. Allen Hynek. He was an astronomer who was invited to act as a scientific consultant to Project Sign in 1948, basically to ensure that any reported sightings didn't have an astronomical origin. He initially believed that the early reports of flying saucers were just a fad generated by unreliable witnesses that would eventually disappear, and he himself went out of his way to provide prosaic explanations to the reports whenever possible. As time went on, however, sightings emerged from very credible witnesses such as pilots and control tower operators that led him to realize that the phenomenon could no longer be glibly dismissed as a transient anomaly. In 1953 he wrote an article in the *Journal of the Optical Society of America* in which he described his learning experience and also criticized those peers whose stance was simply to denigrate witnesses by stating that "ridicule is not a part of the scientific method and the public should not be taught that it is". This article is reproduced at the following link:

http://www.ufoevidence.org/documents/doc1985.htm

Hynek went on to write a best-selling book: The *UFO Experience: A Scientific Inquiry* in 1972 in which he categorized UFO encounters of increasing strangeness as Close Encounters of the First, Second and Third kinds, the last of these being adopted for the title of the blockbuster Spielberg movie on UFOs that premiered in 1977. Hynek also established the U.S. based Center for UFO Studies in 1973, serving as its scientific director until his death in 1986, after which the present director Dr. Mark Rodeghier became incumbent. Hynek acknowledged that he had completely changed his mind on the seriousness of the phenomenon and actively pursued its scientific investigation until he died. Here is a link to a revealing interview he gave in 1985:

http://www.cufon.org/cufon/hynekint.htm

Another early pioneer in the scientific investigation of UFO reports was Dr. James Edward McDonald, a professor of atmospherics physics at the University of Arizona from 1953 until his untimely death in 1971, aged 51. McDonald became interested in UFOs after observing something in the sky with two colleagues in 1954 which they couldn't readily identify. He embarked upon a personal quest to examine as many UFO sighting reports as possible while concentrating on those that were the most difficult to explain in prosaic terms. His professional training meant that he was well qualified to focus on those reports that didn't have a meteorological explanation. After a period of investigation he started giving lectures to highlight what he had discovered and also to plead for a much greater effort by scientists to tackle the UFO question. Of these, perhaps the most important was the one he gave to the American Association for the Advancement of Science (AAAS) meeting in December, 1969 entitled *Science in Default: Twenty-Two Years of Inadequate UFO Investigations.* In his introduction he makes the following comment:

"Charging inadequacy of all past UFO investigations, I speak not only from a background of close study of the past investigations, but also from a background of three years of rather detailed personal research, involving interviews with over five hundred witnesses in selected UFO cases, chiefly in the U.S. In my opinion, the UFO problem, far from being the nonsense problem that it has often been labelled by many scientists, constitutes a problem of extraordinary scientific interest."

He goes on to illustrate his talk with specific examples of high strangeness UFO cases involving very credible witnesses that appear to defy explanation. A full transcription of his lecture can be found here:

http://dewoody.net/ufo/Science_in_Default.html

Other notable examples are a statement he read to the United Nations Outer Space Affairs Group in 1967, as well as a presentation to the American Society of Newspaper Editors in Washington D.C. in the same year. The latter is entitled *"UFOs: Greatest Scientific Problem of our Times?"* These can be accessed using the following links:

http://www.ufoevidence.org/documents/doc1056.htm

http://puhep1.princeton.edu/~mcdonald/JEMcDonald/mcdonald_asne_67.pdf

Unfortunately his evident passion to engender scientific interest in the phenomenon brought him into open conflict with those peers having the opposing view, and his academic standing became threatened. In 1970 McDonald provided evidence to a U.S. Congressional Hearing regarding the potential harm that a proposed supersonic transport plane might cause towards the Earth's ozone layer. One of the congressmen had a personal interest in the development of the plane and countered by pointedly declaring that anyone who 'believes in little green men' was not, in his opinion, a credible witness for the hearing. This attempt at discrediting McDonald's testimony led to laughter among other committee members and left McDonald feeling humiliated. Marital problems then ensued, with his wife seeking a divorce in March, 1971. In June of the same year he committed suicide using a gun, while leaving a note to explain his action. His tragic story is told in an acclaimed book authored by Ann Druffel: *Firestorm: Dr. James E. McDonald's Fight for UFO Science* (2006).

As already mentioned the premier UFO journal in the 1970s was the UK based *Flying Saucer Review*. This was also the organ used by two noted French scientists to publicize sightings they had investigated in their home country as well as to theorize on UFO origins and modes of propulsion. One of these men, Dr. Claude Poher, is still active in the field today. He holds a Ph.D in Astronomy and Astrophysics and was employed for thirty years with the CNES in the area of space research and astronautics. He was also the founder, in 1977, of GEPAN (*Groupe d'Etudes des Phénomènes Aérospatiaux Non-identifiés*), the government funded offshoot that investigated UFO reports and which today continues as GEIPAN referred to earlier. Although now professionally retired he still retains a strong interest in the UFO phenomenon and has even conducted private research on their possible mode of propulsion invoking hypothetical elementary particles which he calls 'Universons', the putative properties of which are described in the following 2012 Applied Physics Research paper:

www.ccsenet.org/journal/index.php/apr/article/download/13522/11293

In this paper the following statement is made:

According to Poher's definition, Universons are elementary momentum carriers, they have no charge, and their speed is c. They exist in the form of a universal flux, and are the source of the mass of particles and gravitation through a specific interaction: absorption – retention – reemission.

Poher has actually patented a method of utilizing these hypothetical carriers to generate a force using superconducting coils which is linked here:

http://www.google.com/patents/US20100251717

A critique of the patent can be accessed here:

http://arxiv.org/ftp/arxiv/papers/1101/1101.2419.pdf

While this experimental work invoking hypothetical Universons is undoubtedly highly controversial, it nevertheless represents an admirable attempt to seek answers to the modus operandi of UFOs. Poher also co-authored a paper entitled *Basic patterns in UFO Observations* which was presented at the AIAA 13th Aerospace Sciences Meeting in January, 1975 and accessed here:

http://www.jacquesvallee.net/bookdocs/AIAA.pdf

The second French contributing scientist to FSR was Dr. Pierre Guerin who died in 2000. He was also strongly in favour of greater participation by scientists in general, but considered that this would be difficult to achieve with the prevailing secrecy over UFOs, and wrote a book entitled *UFOs: the mechanisms of disinformation* (OVNI. Les mécanismes d'une désinformation) that was published just before his death. He reasoned that the policy of secrecy on UFOs would continue because of the likelihood of public unease by its revelation.

Another French scientist who has actively investigated the phenomenon is Jacques Vallee. He studied maths and astrophysics before taking up a position as an astronomer in the Paris Observatory in 1961. He became interested in UFOs after seeing one above his home in 1955 and came into prominence with the publication of his first two books on the subject in the mid-sixties: *Anatomy of a phenomenon: unidentified objects in space – a scientific appraisal* and *Challenge to Science: The UFO Enigma*. In this initial phase he was a strong proponent of the extraterrestrial hypothesis for UFOs. By the late sixties, however, his personal opinion on the phenomenon had changed from it being a manifestation of extraterrestrial visitation to one involving an inter-dimensional component, such that it was an interaction from a co-existing domain. He postulated that many so-called 'paranormal phenomena' and other unusual experiences were related to UFOs and wrote a book in 1969 exploring the idea called *Passport to Magonia: From Folklore to Flying Saucers*. He has since written many other books along the same theme, with the result that he freely admits to being a 'heretic amongst heretics' concerning his views. In 1990 he penned an article entitled: *Five Arguments Against the Extraterrestrial Origin of UFOs* which can be viewed here:

http://www.jacquesvallee.net/bookdocs/arguments.pdf

One other French scientist with an abiding interest in UFOs is Jean-Pierre Petit who was a senior researcher at the French CNRS and a pioneer in magnetohydrodynamics. Although now retired he recently published a Kindle book entitled: *UFOs and Science: What has been Scientifically Discovered* (OVNIS et science: Ce qu'ont découvert les scientifiques) and has his own website on the same theme:

http://www.ufo-science.com

While the contribution of French scientists to the investigation of

UFOs is apparent, that from British scientists has been sadly lacking. The UK science establishment has unfortunately followed a position of disinterested aloofness regarding the phenomenon. As recently as September 2012 the Astronomer Royal, Lord Martin Rees, proclaimed that "only kooks see UFOs" without providing any hint whatsoever of any scientifically based references to justify such a claim:

http://www.huffingtonpost.com/2012/09/19/lord-martin-rees-aliens-ufos_n_1892005.html

A week after the above article appeared the same Huffington Post published a rebuttal by providing survey evidence from professional astronomers who had reported seeing unidentified phenomenon in the sky at a greater rate than the public at large!

http://www.huffingtonpost.com/dan-mack/astronomers-ufo_b_1901480.html

One of the surveys cited was conducted by Dr. Peter A. Sturrock, a British born astrophysicist who recently retired as emeritus professor of applied physics at Stanford University. His survey can be accessed here:

http://www.scientificexploration.org/journal/jse_08_3_sturrock.pdf

Sturrock is, to my knowledge, the only British scientist of repute who has been actively involved in studying the UFO phenomenon. He came into it in the early 1970s almost by accident, having hired Jacques Vallee for a research project because of his experience in both computers and astrophysics. On learning of Vallee's many books on UFOs he felt obliged to read some of them. This sparked an interest which then led him to study the Condon Report. As other scientists had already discovered, Sturrock concluded that Condon's negative conclusions on the further investigation of UFOs wasn't supported by the evidence that was presented in the actual report. He therefore initiated his own investigation by finding out what other scientists might have seen or felt regarding the UFO controversy, leading to the aforementioned surveys. One of the results to come out from these surveys was that many of its respondents wished to see UFOs discussed in scientific journals. Sturrock therefore became instrumental in establishing in 1982 the Society for Scientific Exploration, which would provide a scientific forum for the discussion of UFOs and other topics neglected by mainstream science. Its main journal is the *Journal of Scientific Exploration* which has been published since 1987.

In October, 1997 Sturrock also organized a four day workshop to review the physical evidence associated with UFOs. This took place in Tarrytown, New York, and had a review panel comprising of nine scientists of diverse experience and interest. The Proceedings of the workshop can be read here:

http://www.scientificexploration.org/journal/jse_12_2_sturrock.pdf

He then wrote a book on the same topic entitled: *The UFO Enigma: A New Review of the Physical Evidence* which was published in 2000. More recently he has had published a second book, the title of which refers to his 'double' life as an orthodox physicist who has also probed esoteric phenomena: *A Tale of Two Sciences: Memoirs of a Dissident Scientist* (2009). A review of this can be accessed here:

http://www.thiemeworks.com/a-tale-of-two-sciences-memoirs-of-a-dissident-scientist-by-peter-a-sturrock/

The establishment by Hynek of the Centre for UFO Studies (CUFOS) led to the latter becoming a focal point for the investigation of UFOs in the U.S. After Hynek's death its direction was taken over by Dr. Mark Rodeghier who has a first degree in astrophysics and both an MA and Ph.D in sociology, the latter arising through studying scientists' attitudes to controversial research. An interview that reveals how he became involved with

UFOs and CUFOS can be read here:

http://www.cufos.org/bio_markrod.html

Rodeghier has written many articles on aspects of UFO research and history, a notable one being the involvement of the CIA in its alleged cover-up:

http://www.cufos.org/IUR_article3.html

Here's another on U.S. government document accessibility on the internet:

http://www.cufos.org/UFO_Documents_internet.html

Perhaps Rodeghier's most important research contribution to date is the cataloguing and analysis of 441 vehicle interference cases in which the close proximity of a UFO often causes the temporary malfunction of the ignition system of the vehicle. An important finding from this study is that diesel engines appear to be much less prone to such malfunction, a summary of which findings can be read in the Proceedings of the Sturrock four day workshop on physical evidence referred to above (p.195). It is anticipated that the vehicle study will soon be accessible in full on the CUFOS website.

As the involvement of scientists with CUFOS increased in the 1970s and 1980s it became evident that a journal devoted to disseminating scientists' views and research material to their peers was merited. Thus the *Journal of UFO Studies* was launched. The first of three volumes was published in 1979, concluding with the third in 1983. The list of titles for the papers presented in these first three volumes can be accessed here:

http://www.cufos.org/jufosold.html

There followed an interval before publishing was restarted in 1989, continuing until 2003. This 'new series' of eight volumes, the first of which included my initial paper on the Delphos chemical analysis, again comprised of peer reviewed articles devoted to the scientific study of the UFO phenomenon:

http://www.cufos.org/jufosnew.html

The editor of the 'new series' of JUFOS was Dr. Michael D. Swords, a board member of CUFOS and currently Professor Emeritus of Environmental Studies and Natural Sciences at Western Michigan University. Like other interested scientists before him, Swords had his own sighting in 1959, when he observed with others an unusual object in the sky, thus initiating his curiosity. He has since written many articles on the phenomenon, an early example of which appeared in the same JUFOS volume as my Delphos paper, being entitled *Science and the Extraterrestrial Hypothesis in Ufology*. This can be read here:

http://www.thiemeworks.com/michael-swords-on-the-extraterrestrial-hypothesis/

Sword's in-depth exposition of the Colorado University UFO Project of 1967 (which led to the publication of the flawed Condon Report) can be viewed here:

http://www.cufos.org/Condon_The_Scientific_Study_of_UFOss.pdf

Other articles of his in review form have appeared in the Journal of Scientific Exploration, such as *A Guide to UFO Research* which is linked here:

http://www.scientificexploration.org/journal/jse_07_1_swords.pdf

This was followed by one entitled *Could ETs Breathe Our Air?*:

http://www.scientificexploration.org/journal/jse_09_3_swords.pdf

Mention has already been made of the comment by Astronomer Royal Martin Rees that 'only kooks' see UFOs. Rees might be interested to learn that Clyde Tombaugh, the famous astronomer who discovered the planet Pluto, experienced seeing them on no less than *three* occasions, as discussed by Swords in this article:

http://www.scientificexploration.org/journal/jse_13_4_swords.pdf

And here's Sword's overview - wittily elaborated - of the phenomenon itself, entitled *Ufology: What Have We Learned?*:

http://www.scientificexploration.org/journal/jse_20_4_swords.pdf

Doctor Swords has also had published two books on the subject: *Grassroots UFOs: Case Reports from the Center for UFO Studies* (2011), and *UFOs and Government: A Historical Inquiry* (2012), the second of which he was the primary author regarding the U.S. side of the inquiry, with others contributing on countries such as Sweden, Australia, France and Spain. The book is widely considered to be the most complete and definitive account on how governments around the world have dealt with the UFO phenomenon since its inception in the 1940s.

Swords is also responsible for the running of an internet blog on (mainly) the UFO topic:

http://thebiggeststudy.blogspot.co.uk/

Another scientist who has delved deeply into the UFO question is Stanton Friedman, a former nuclear physicist, whose interest was sparked off by reading a book on Project Bluebook in 1958. At the time he was undertaking research on the possibility of using nuclear power to propel aircraft and was curious to learn whether the reports of fast moving flying discs might implicate such a power source. As is usual for a genuinely curious mind, the more he read on the topic the more he became convinced, which ultimately lead him to conduct his own research. After fourteen years of employment within various big name companies including McDonnell-Douglas, the nuclear power programmes he was involved in were eventually cancelled, and he made a move into lecturing about UFOs as a profession, focusing on venues such as universities and colleges, alongside his other consultative activities.

Friedman's original research delved into the Roswell incident which led him to co-author a book with Don Berliner (an aviation/science writer) entitled *Crash at Corona: The Definitive Study of the Roswell Incident* (1992). He has done a great deal of other research into sourcing classified archival material using the Freedom of Information Act (FOIA) to retrieve many CIA documents relating to UFOs, some of which had large sections blacked out to render them unreadable. Because of their existence he considers these documents to constitute strong circumstantial evidence for there being a massive U.S. cover-up of UFO evidence which he calls the 'Cosmic Watergate'. He has written several books on the UFO theme, two of which I would recommend for their discussion of how certain scientifically trained 'experts' but having little actual knowledge of the subject matter have not only misrepresented the UFO conundrum but, lamentably, have also led to ill-considered pronouncements within other unrelated areas of research that have adversely affected and even delayed their successful conclusion. The books in question are *Flying Saucers and Science* (2008) by Friedman himself, and *Science was Wrong* (2010) which Friedman co-authored with Kathleen Marden. He also has his own website:

http://www.stantonfriedman.com/

Other U.S. based scientists who have been intimately involved in UFO research include Dr. Bruce Maccabee and Dr. Richard F. Haines. The former is an optical physicist who started his research in the late 1960s and has used his expertise to analyze photographic evidence for UFOs. Although there are countless examples of such, their provenance is very often suspect, and the only way alleged photographs of UFOs can be checked for their authenticity is by undertaking proper analysis using specialist equipment which Maccabee has access to. He has also been involved in document retrieval from government archives, particularly those involving the FBI and UFOs. Maccabee has his own website which is easily searched to locate his extensive research files:

http://brumac.8k.com/

Here, for example, is the first part (of three) of his analysis of the famous 1950 McMinnville Trent Farm photographs:

http://brumac.8k.com/trent1.html

Dr. Haines was formerly a NASA scientist and has, for the past thirty years, been involved in compiling and researching UFO sightings reported mainly by professional pilots along with any associated radar and photographic evidence. He has accumulated close to 3,500 reports so far. In 1999 he established a dedicated centre for the reporting of such events called NARCAP (short for National Aviation Reporting Center on Anomalous Phenomena) which is linked here:

http://www.narcap.org/index.html

A principle focus for NARCAP is air safety, since there have been numerous instances of aircraft being approached by a UFO with the pilot having to take corrective action to avoid a possible collision. On the 'About NARCAP' page of the website the following statement - referring to UAP (unidentified aerial phenomena) in place of UFO - is made:

Dr. Richard F. Haines has compiled a catalogue of over 3400 aviation related UAP cases. He has conducted a comprehensive review of UAP reports by U.S. air traffic controllers and pilots from the past 50 years. A result of this effort is Richard's paper "Aviation Safety in America- A Previously Neglected Factor". It contains analyses of over one hundred reports of UAP involved in near misses, close pacing, disrupted avionics, and collisions. These events either occurred to US aviation professionals on domestic or foreign flights, or foreign aircrews operating in US airspace. This document includes Dr. Haines' recommendations for addressing these issues. This paper is not copyrighted and is being widely distributed throughout the aviation industry.

There follows the statement below concerning confidentiality and a request for pilots to come forward with their reports while ensuring that this would be maintained:

It is our hope that aviation professionals will recognize the importance of this work and contact us with their reports of encounters with UAP. Often, though not always, reporters are concerned about their confidentiality. We are not associated with the FAA or other government agencies, or the airlines. We have modelled our program after NASA's Aviation Safety Reporting System (ASRS). With regards to employers, the FAA, and the media, we have a process in place to ensure that confidentiality is protected.

We are primarily interested in reports from pilots, air traffic controllers and radar operators. However, anyone who witnesses UAP appearing to represent a threat to aviation safety may contact us.

Investigations:

We strongly suspect that UAP adversely affect avionics and aircrews. We are interested in their impact on aviation safety and will carefully examine each incident where safety may have been compromised. We will conduct thorough investigations of UAP reports based on aviation industry standards. Our Technical Advisors have extensive aviation and aeronautic experience and will follow established patterns of aviation safety investigations and reporting.

Our International Science Advisors represent a cross section of disciplines, from geophysicists and research psychologists to meteorologists and astrophysicists. We expect that they will be quite helpful in evaluating cases and offering research of their own for review.

Analysis, theories or findings related to UAP will be posted and open, critical analysis by peer review will be encouraged. Additionally, we are encouraging written submissions by credentialed parties who wish to present their material for peer review.

We are very interested in networking with aviation safety groups in particular, as well as the aviation community in general. Collaborative efforts are encouraged.

The NARCAP website includes a list of 'Technical Reports' that have been issued by the centre. Two of these are linked below. The first describes an encounter with a UFO by a Jumbo Jet shortly after its take off from Los Angeles Airport in September, 1996. Both pilot and co-pilot observed in front of them a pair of bright lights similar in appearance to aircraft landing lights but directed towards them and keeping pace with their own jet, such that if they had been landing lights it would have meant that the plane in front was flying backwards! After about ten seconds the object shot off at a slight elevation towards the left before disappearing in a matter of seconds leaving the witnesses awed by the speed of its departure, described as performing 'instant' acceleration:

http://www.narcap.org/files/narcap_UC_MainText_9-27-96.pdf

The second report is actually a catalogue of 1,305 military, airliner and private pilot reports of UFOs from the period 1916 to 2000. This catalogue should serve as an eye opener to anyone who doesn't realize that pilots do have encounters with UFOs but whose stories do not, for one reason or another, end up being reported by the general media.

http://www.narcap.org/files/narcap_revised_tr-4.pdf

A final entry into this list is someone I became aware of only relatively recently. The astronomer and meteorologist in question is Dr. Eamonn Ansbro who has been active in SETV (i.e. the Search for Extraterrestrial Visitation), which, curiously, is an offshoot of SETI that is slowly gaining scientific respectability. It concerns itself with the possibility that alien probes are already present in Earth's neighbourhood and may therefore be monitoring us from this vantage point. SETV scientists are therefore keen to develop and deploy specific technology to discover if such probes might exist:

http://www.carlotto.us/newfrontiersinscience/Papers/v01n04a/v01n04a.pdf

http://www.daviddarling.info/encyclopedia/S/SETV.html

Ansbro also happens to be a professional instrument maker, and founded OptiGlas Ireland Ltd in Dublin in 1981 to manufacture optical equipment to high specifications. He has used this facility to furnish the Kingsland Observatory in County Roscommon with telescopes fitted with CCD cameras in order to constantly monitor the skies for alien probes. The observatory now boasts a whole sky surveillance technique that can recognise and track targets while triggering video recordings at the same time. His work has been acknowledged by the European Space Agency (ESA) and the group of scientists he works with is known as OSETI (i.e. Optical SETI). He is also part of the UK SETI network and gave a talk entitled *Searches* for *Robotic Probes in the Solar System* in their July 2013 Symposium, the programme for which can be viewed here:

http://star-www.st-and.ac.uk/~ap22/setinam2013.html

The actual paper that Ansbro delivered at the symposium is linked here:

http://www.seti.ac.uk/dir_setinam2013/slides/slides_NAM2013_eamonn_ansbro.pdf

And here's a paper co-authored by Ansbro entitled *SETV: Opportunity for European Initiative in the Search for Extraterrestrial Intelligence:*

http://www.europeanufosurvey.com/en/setv.php

Ansbro's efforts to promote the investigation of near-Earth alien probes and, by inference, UFOs is summarized in this feature article that appeared in the Irish Times:

http://fionolameredith.co.uk/features/eamonn_ansbro.html

It remains to be seen whether these on-going SETV and OSETI activities by Ansbro and his colleagues might ultimately lead to a serious scientific investigation of UFOs.

There have, of course, been many other quotes attributed to leading scientists on the UFO phenomenon in the past, a selection of which can be found here:

http://www.ufoevidence.org/documents/doc1744.htm

http://www.ufoevidence.org/topics/science.htm

The foregoing should serve to illustrate the fact that scientists who are initially open-minded on the UFO conundrum and then go on to study it will generally adopt the view that it merits very serious investigation rather than the aloofness so far displayed by the scientific community as a whole.

Part 3. The analysis of the Delphos ring soil and its implications.

Although my full scientific report is attached at the end of this book (see Appendix), to satisfy those who don't necessarily wish to plough through the detail of the science I have included below a part re-worded summary that will explain the results and its implications in more general terms.

The event occurred on November 2, 1971, at Delphos, Kansas, at 7:00 p.m. local time. A sixteen-year old boy, Ronald Johnson, was tending his sheep at his father's farm when he heard a rumbling noise and then stepped out of a small shed to see an illuminated object hovering beneath a tree about 75 ft. away from him. The object had an estimated diameter of 9 ft. and appeared to be about 10 ft. high (Fig. 1). The rumbling noise was likened to that of a washing machine that vibrates when out of balance.

The boy described the object as multicoloured with blue, red and orange glows about its entire surface as it hovered 2-5 feet off the ground. He also observed a bright glow between the object and the ground, as though a shimmering material was falling from the object. The boy said that it hurt his eyes when looking directly at the object and for several days after the incident his eyes were painful and he suffered headaches. After some minutes the object began to move off passing over a nearby shed by about 4ft at which point it emitted a high pitched sound like that of a jet aircraft (Fig. 2). Ronald briefly lost his vision while observing the illuminated object, but after his sight returned he looked into the sky and could see the bright light receding into the distance. Ronald ran back to the farm and told his parents what he had seen. They didn't believe him at first and he became agitated. They all then went out to observe in the southern sky a bright luminescent object receding into the distance bearing "the colour of an arc-welder." It was estimated to be at least half the size of the full moon, which could also be seen in the south-eastern sky at the

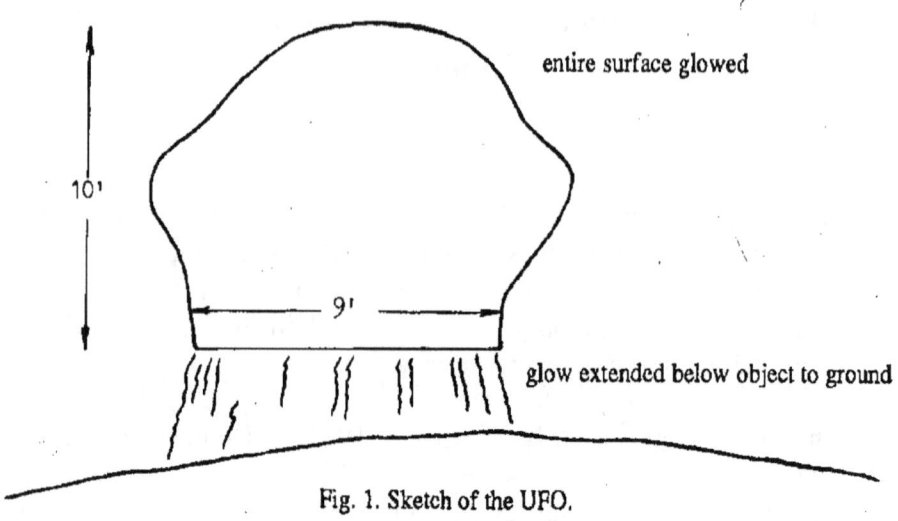

Fig. 1. Sketch of the UFO.

same time. Although this was the last time that the family members saw the object another witness Lester Ernsbarger, a reserve police officer in Minneapolis (ten miles south of Delphos), called in to his base to report a bright light in the sky which he had observed at 7.30pm in the northern sky from his position, thus providing

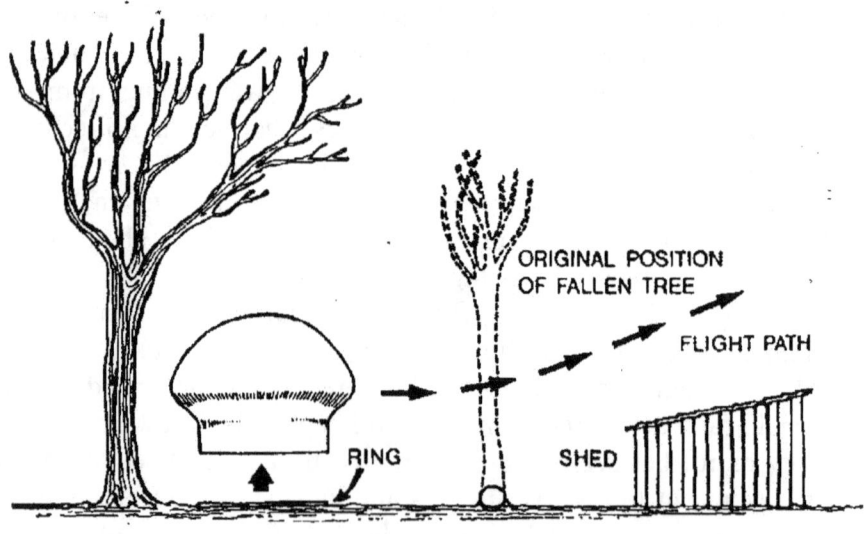

Fig. 2: Initial hovering and departure of UFO.

possible independent corroboration for the presence of the illuminated object (Fig. 3). The Johnson family then walked to the area where the object had originally been observed. As they approached they saw a ring of soil over which the UFO had hovered glowing in the dark. Portions of the nearby trees reportedly also glowed. The glowing ring so impressed Mrs. Johnson that she ran back to the house to fetch a Polaroid 104 camera with which she took a photograph of the effect (Fig.4). No flash attachment was used to take the photograph, as the light from the ring was described as bright enough to read a newspaper by. She had only one exposure remaining in the camera and could see the illuminated ring through the view finder well enough to get the camera positioned. The family claimed that the glow from the ring lasted long enough to be visible the following night.

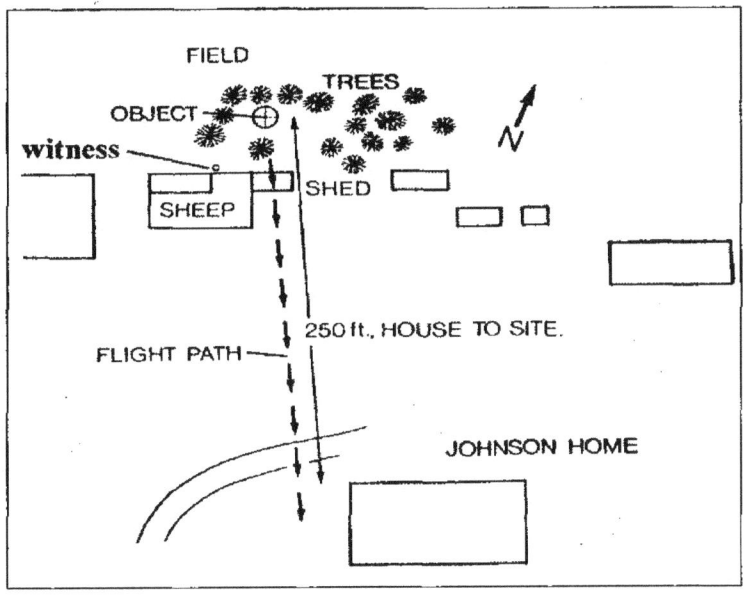

Fig. 3: Site plan showing UFO and farm area

The witnesses proceeded to touch the ring, which they described as having a cool, slick, crust-like texture, as if the soil was "crystallized" as well as being moist and pitted with tiny craters or holes on the surface. They noticed an unfamiliar odour on their fingers and immediately experienced a numbing effect similar to that of a local anaesthetic which took several days or weeks to wear off. The Johnson family reported the alleged event to the local sheriff, who took samples and a photograph of the ring the next day (Fig. 5), and to a newspaper reporter.

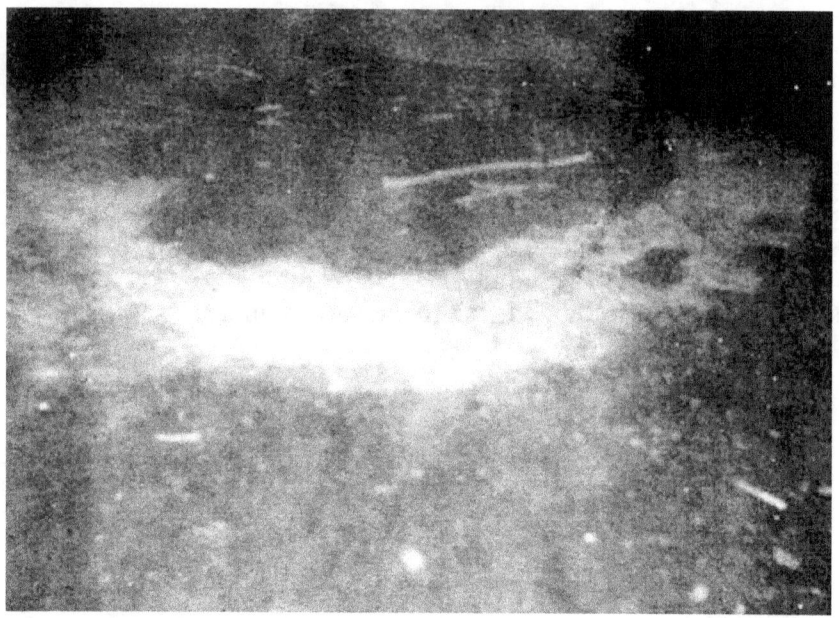

Fig. 4: Polaroid photo of glowing ring (taken without a flash attachment)

Ted Phillips arrived on the scene a month later to interview the family and measure and sample the ring soil. His initial report (Phillips 1972a, 1972b) emphasized the credibility of the witnesses and the puzzling nature of the ring which now displayed pronounced water repellent character, so much so that it was still covered by un-melted snow, in contrast to the surrounding muddy environment. This water repellent property extended in places to 14 inches below the surface and lasted for some months afterwards. The ring had an unaffected portion at its north-western edge, while the whole was elongated towards the south-east (Fig.6). A check of weather conditions prevailing at the time showed wind direction and speed to be 290° and 8 mph respectively, suggesting a possible wind spreading effect.

The Compelling Scientific Evidence for UFOs

Fig. 5: Photograph of ring taken 19 hours after the alleged event.

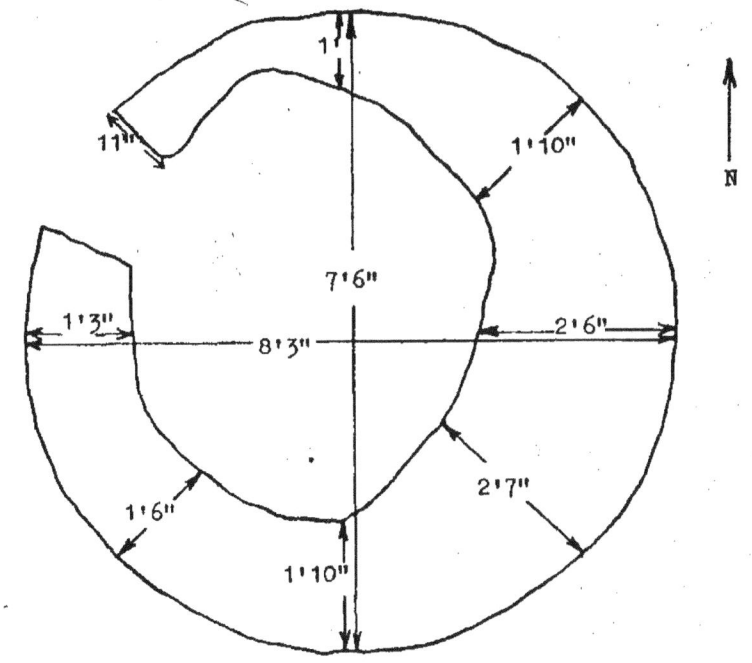

Fig. 6: Dimensions of ring soil.

Because of the obvious importance of the case samples of ring soil were duly sent to various laboratories in the U.S. to analyze. Unfortunately, because of lack of sufficient funding and resources none of these laboratories were able to offer any clues as to the nature of the chemical changes that had arisen to cause the ring soil absorption anomalies. This dearth of information was the reason why I requested the material to examine for myself while at Nottingham University.

On receiving the ring soil I immediately checked for its alleged water repellent nature and was stunned to discover how marked this was. Placing water onto the affected soil was very like placing it onto a glass surface, with the water spontaneously forming into droplets sitting on the surface. This behaviour was even apparent from samples taken some inches beneath the surface! With my curiosity piqued I took a spatula and tried to force the material to mix with the water, which it duly did to provide a homogeneous suspension.

Centrifugation of the suspension afforded a reddish brown solution sitting above the separated soil matter. When this solution was irradiated using a UV lamp it was found to fluoresce which a chemist would know was consistent with the claim that the soil had been observed to glow on the night of the event. I then advised my academic supervisor Dr. Barrie Bycroft (now a retired professor; contact details of whom can be supplied on request) of the case particulars and described to him the unusual characteristics of the affected soil. He became intrigued and referred me to the university expert on chemiluminescence Dr. Frank Palmer (now deceased). When I went to see Dr. Palmer in his office he initially had a smug grin on his face, clearly having heard that this involved a UFO report. I then told him of the case particulars and odd nature of the soil and showed him the photograph of the ring which I explained had originally glowed. He then observed for himself the fluorescence of the ring soil extract, at which point his facial expression transformed into one that could best be described as astonished bewilderment. After recovering from his initial shock he agreed to help analyze the soil, and was personally responsible for recording the fluorescence spectra presented in my final report.

The reader should be reminded here of the sequence leading up to the formation of the glowing ring soil. The closest witness to the UFO, Ronald, heard a rumbling noise and stepped out of a shed to view the illuminated UFO with blue, red, and orange glows about its surface. However, he also observed a simultaneous glow *between* the object and the ground while it hovered underneath the tree. And then, after the object departed, the family ventured to the area and saw a glowing ring over which the UFO had been positioned. So we have *three* separate glows observed which would appear to be linked in view of the short time lapse between them.

While inspecting the red-brown solution derived from the ring soil I discovered that it foamed on shaking, similar to how a soapy solution would behave. This suggested that the compound present might be in the form of an alkali metal salt of an organic carboxylic acid. I therefore treated the solution with an aqueous solution of silver nitrate which succeeded in precipitating out the silver salt of the suspected acid. This was centrifuged off, washed and dried. An infra-red absorption spectrum of the silver salt showed the presence of strong hydroxyl and carboxyl groups which conferred strong water solubility to the compound and - ironically - it was this very same property which paradoxically gave rise to the water repellent nature of the soil, since these groups were closely bound up to the similar water adherent groups of the soil resulting in a powerful masking effect of the latter (the replication of this repellency using a water soluble anthraquinone derivative as a test substance is described in the Appendix).

The isolated silver salt was then reacted with methyl iodide in acetone to generate the corresponding methyl ester of the carboxylic acid. The compound in the ring soil was now in a form that could be readily analyzed using a method called thin layer chromatography. This technique enables a mixture of chemical components to be separated into individual ones and therefore function as a purification method. On completion of the separation a major yellow band was obtained which under UV light appeared dark. On leaving to stand for some minutes the yellow colour faded (i.e. decolourized) while at the same time its darkness as viewed under UV light was substituted by a highly fluorescent bluish-white appearance. When the chromatogram was covered with a glass slide to reduce exposure to air these changes did not occur. This was an important observation, because it meant that the main component in the mixture was air sensitive and readily oxidized to give a fluorescent compound, and this was precisely the sort of property that would be expected for a potentially chemiluminescent oxidation reaction. The latter term refers to a chemical reaction that results in the emission of light! This finding can now be used to address the glows observed by the witnesses on the night of the event. In simple terms the ring soil contains an organic compound which potentially could emit light when exposed to air. It should be emphasized that I did attempt but was unable to demonstrate actual light emission from crude aqueous extracts of the ring soil, but this is likely to have been due to the appreciable degradation of the active component in the soil during the six years that had elapsed before my eventual analysis.

The observation that the UFO was emitting light before leaving a glowing ring in the ground strongly suggests that the soil compound was responsible for the former. The model that I propose for the ring's formation is the one shown in Fig 7. This depicts the UFO containing within its peripheral walls an aqueous solution of the chemiluminescent compound which was being employed to generate light about its surface. This surface would necessarily have to be constructed from a transparent material in order for the light to be observed and would also have to be permeable to oxygen to ensure the latter could enter and react with the solution. The symbol $h\nu$ is a scientific representation of the light energy being generated (which is proportional to its frequency ν).

During Ronald's observation there was also the glow *between* the object and the ground. This could conveniently be explained as being caused by a simultaneous deposition of the chemi-luminescent solution which is shown as a spray. As this was dropped it would immediately react with the oxygen in the air and thereby create the perception of a glow beneath the object. This postulated spray would also explain why the ring is elongated towards the wind direction on the night of the event as well as lead to the 'pitted' sensation noted by the family on touching the ring soil surface. The white colouration of the ring soil as seen in the photograph would be due to a layer of the final oxidized material (after the luminescence had subsided) sitting on top of the ring soil since the active compound decolourizes during the oxidation. The spraying theory would also imply that the surface of the ring soil should have been moist to the touch. In fact, when Phillips questioned the witnesses this was indeed reported to be the case by them along with the numbing effect that resulted when the surface was touched. The numbing effect is akin to local anaesthesia and indicates the presence of a compound having dual water attracting and repelling groups - which again is corroborated by the fact that the ring soil became water repellent.

The notion that the external wall material should be transparent while also allowing oxygen to enter is not an impossible one. There is already in commercial use a clear plastic material called polymethylpentene which has appreciable gas permeability and is used to make films and coatings for substances that require such packaging. To enable oxygen to enter the wall there would need to be generated a partial vacuum to draw the gas in. And the model allows for this because the proposed ejection of part of the solution as a spray from beneath the UFO would create such a vacuum. As to the various surface colours of the UFO observed by Ronald – i.e. blue, red and orange – even these can be explained by the analysis data. As has already been mentioned the natural colour of the solution impregnating the soil is red. However, the fluorescence of the oxidized compound is bluish white when observed under UV light, and this would equate to the colour of the emitted light. Since aerial oxygen needs to be absorbed through the transparent surface of the UFO for light emission to occur, one can imagine a scenario whereby a non-uniform absorption would produce the blue light in certain areas only, while the remainder of the surface would remain as red or orange being the illuminated natural colour variation observed for the un-oxidized solution. Once the object had departed the family described it as bearing the 'colour of an arc welder'. This would be consistent with the - now uniform - absorption of oxygen about the surface giving rise to an overall surface emission of the blue light. One can speculate why the UFO would choose to employ a luminescence technology such as this. Certainly many UFOs are frequently described as being luminous, and this might be one method of achieving this.

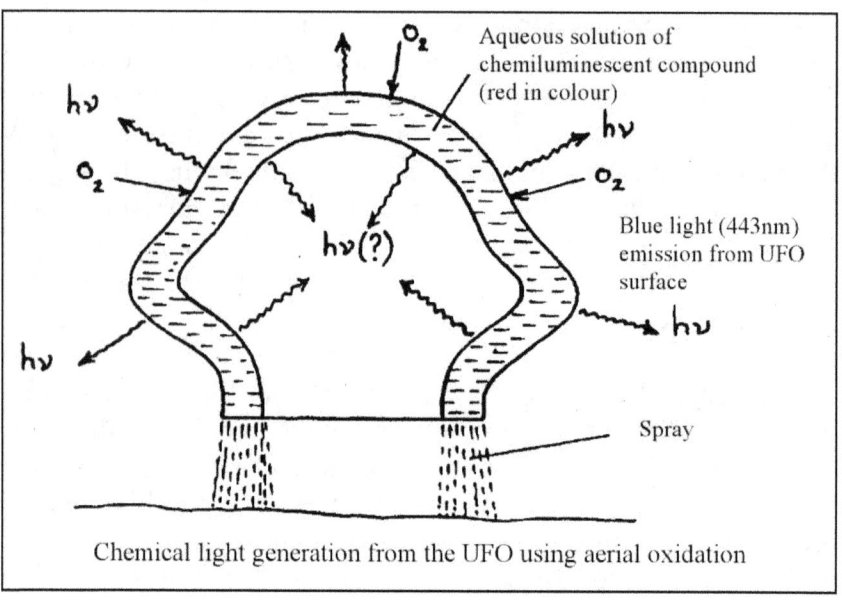

Fig. 7: Possible mode of light emission from the UFO

The technology has obvious drawbacks, however, such as the fact that it is chemically based, and the light emitting material will ultimately be spent and need re-charging. It will also mean that a relatively heavy aqueous solution will be carried by the UFO, which would have to be taken into account regarding whatever propulsion method is being used by the craft. Might this load be the reason why some of the material was dumped into the ground by the UFO? And performing this while under a tree would also ensure that the freshly deposited luminous material wouldn't be seen from a passing aircraft after the UFO had departed.

The luminosity of UFOs may be being employed as a way of engendering an enigmatic quality for terrestrial observers to be puzzled by. However, it might also be considered as an 'ultimate' light source which would uniformly illuminate not only the outside of the UFO *but also the inside*, as indicated in Figure 7. Intriguingly, this latter possibility can now be seen to tie in with the experiences of alleged abductees who claim to have been taken on board UFOs and then be bathed in a light that 'comes from everywhere'!

Comparison of the Delphos ring with naturally occurring fungal 'fairy rings'

It has been mentioned by Vallee in some of his books that inspection of the Delphos ring soil had revealed the presence of fungal material that he identified as Nocardia. For example in *Forbidden Science* the following statement is made: "An analysis of the white matter in the glowing ring left after the sighting disclosed the presence of a Nocardia fungus that seemed to have been stimulated by an unknown radiation." One might assume from this that the ring might therefore simply be a product of natural fungal growth that is sometimes seen in forested or grassy soils and referred to as 'fairy rings'. The latter normally start from a central point of infection before moving outwards with time to afford a slowly enlarging ring shape. Frequently these rings are characterized with mushrooms visibly growing on their surfaces. When I inspected the ring soil I observed very small amounts of fungal matter in the ring soil which are likely to have been the same material that Vallee reported. However, these amounts were miniscule in comparison to the large amount of water soluble organic substance impregnating the soil. The overwhelming impression I gained during examination was that the fungal material had proliferated *because* of the enriched organic content of the soil, and not vice versa! Since it took a month for the investigator, Ted Philips, to arrive at the scene and then sample the soil, this would have allowed sufficient time for any indigenous micro-organisms to multiply and become visible during that period. One should also bear in mind that the soil was *not* the type usually noted for 'fairy ring' formation, and the ring did *not* then grow outwards with time as would be expected for a genuine fungal ring.

In the Foreword to this book, Michael Swords refers to the renowned expert on fungal diseases, Dr. Hubert Lechevalier of Rutgers who also inspected the soil and concluded that the ring, "although containing some fungal elements as almost all soils do, was NOT the product of such an agency" Finally, it should be apparent that the elongation of the Delphos ring towards the wind direction on the night of the event would be extremely difficult to explain within the context of a natural fungal 'fairy ring' explanation!

Comparison of Delphos with other physical trace cases

As far as I know this case is unique in having a multiple witness report of a UFO that is alleged to have left behind an analyzable chemical residue. There are, of course, many examples of physical evidence apparently left by UFOs while close to or on the ground, but these are mainly of depressions or scorch marks which are of limited value as far as analysis is concerned. And with many of these cases the opportunity for carrying out a proper scientific study had been unfortunately lost anyway. In the original 1972 case report for Delphos there is listed a number of similar cases at the report's conclusion which readers can view and compare for themselves:

FSR-CH 1972 N 9.pdf - noufors.com (use Google search to find)

A case that is often lauded when referring to physical evidence is the Trans-en-Provence event which occurred in France on January 8th, 1981. The single witness was Renato Nicolai, a farmer who was working on his property when he heard a whistling sound and observed an eight foot diameter metallic saucer shaped object land about 50 yards from him. Shortly afterwards it flew off, leaving behind a compressed area with associated scorch marks. An analysis of the area was subsequently conducted by GEIPAN which concluded from the compression depth that a weight of 4 - 5 tons had settled on the ground, while a temperature of between 300 - 600 degrees Celsius was considered to have caused the scorching.

Nearby plants had also allegedly been affected by heat through discolouration. Further details on the case and its analysis can be found here:

http://www.scientificexploration.org/journal/jse_04_1_velasco.pdf

http://www.ufoevidence.org/cases/case110.htm

The GEIPAN website, part of the French space agency CNES, can be viewed here:

http://www.geipan.fr/

As with many other physical trace cases the Trans-en-Provence event highlights the usual problem that is encountered if any subsequent analysis is attempted. Even though marks or indentations have apparently been left behind by the reported UFO there is little that can be concluded about what these effects actually mean or represent! Consequently any attempt at extrapolating from the results some kind of proof that the UFO actually existed ultimately fails. And the sceptics can therefore maintain their mantra of: 'there is no good evidence for UFOs'! I would contend that the Delphos case breaks this impasse, because there are several corroborating factors which combine to make a very powerful case for the existence of the UFO:

1. The close-up observation of the glowing hovering UFO with a simultaneous glow seen between the object and the ground.

2. The multiple witness observation of the glowing object seen in the sky after departing.

3. The multiple witness observation of the glowing ring allegedly left behind by the UFO after it departed.

4. The Polaroid photograph taken of the glowing ring within minutes of its alleged formation.

5. The photograph taken by the local sheriff the next morning of the white coloured 8ft diameter ring reportedly left behind by the UFO.

6. Measurements subsequently taken of the ring soil which reveal a clear skewing towards the wind direction that prevailed at the time of the sighting.

7. Subsequent chemical analysis of the ring soil which revealed the presence of a highly water-soluble and potentially light emitting compound.

8. The facile ability of using this data to formulate a model that explains all the above effects invoking a hitherto unknown technology and suggesting in-turn the presence of an unknown aerial object.

9. For those readers wishing to gain a wider knowledge of the kinds of UFO physical trace cases that are reported I would refer them to the three links below. The last of these details the famous Westall Incident of Australia in April 1966 in which over 200 students and staff from two schools watched a silvery disc hover and then land in a paddock before flying off to leave a circle of discoloured grass behind.

http://www.ufoevidence.org/topics/physicaltracecases.htm

http://www.ufoevidence.org/Cases/CaseView.asp?section=PhysicalTrace

http://www.ufoevidence.org/cases/case591.htm

Ufoevidence.org is noted as having one of the largest compilations of UFO reports of *any* category that can be readily accessed on the web. In addition, the Center for UFO Studies website posts a good introduction on UFO evidence which can be accessed here:

http://www.cufos.org/FAQ_English_P2.html

Part 4. Aftermath of Delphos ring soil analysis

Once I had collated all the analysis data I naturally considered writing a report detailing my findings and its conclusions. I initially questioned whether I should be doing this at all, since the report might conceivably lead to an opening of a Pandora's Box as far as UFOs were concerned. However, my scientific instincts overrode such concerns, and I therefore wrote and sent the report to CUFOS in Chicago where it was initially read by Rodeghier and Clark who both considered it important enough to propose its publication in the Journal of UFO Studies (New Series). It duly appeared in the 1989 issue (under the editorship of Dr Michael Swords), while a summary was also published in the International UFO Reporter (edited by Clark) in two parts (Volume 12, Nos. 1 and 3). In private correspondence from that period Clark believed my report to be a 'major contribution' to UFO research, while Ted Phillips found my conclusions to be of 'great interest' and believed that the recently deceased founder of CUFOS, Dr J. Allen Hynek would have concluded the same.

Phillips also alluded to the formation of a 'Coalition for Delphos', which I took to mean a grouping of UFO study organizations that would take the case to a logical conclusion. Surprisingly, however, there was very little comeback from the publication of my report. A couple of follow-up articles on Delphos appeared in JUFOS (one written by myself) but there was little in the way of discussion within ufological circles of my report or its conclusions.

Phyllis Budinger's analysis of the ring soil

Then in 1990 Phillips contacted me to say that he had approached another analyst, Phyllis Budinger, who would be able to do further analysis on the Delphos soil and introduced me to her by email. There followed a period of tri-party discussion by email as Budinger got to handle and become familiar with the soil. I pointed out that she had to forcibly wet the ring soil with water to extract out the compound. She then proceeded to reproduce my isolation procedure in order to obtain the same silver salt that I had isolated. However, what she was not able to do was then react this silver salt with methyl iodide in order to prepare the methyl ester and isolate the active component from the ring soil using thin layer chromatography. This last procedure required a synthetic step which, for a pure analyst, may have been difficult to perform without suitable resources.

Budinger's analysis corroborated my own as far as it went, but without the additional thin layer chromatography data she wasn't able to conclude in terms of an air sensitive, light emitting compound as I had done. What she found from her partial analysis was a water soluble organic compound that had degraded in the soil to give a humate-like substance, which she called 'fulvic acid'. Although this name suggests a naturally occurring material, Budinger nevertheless thought it was likely to have been deposited by the UFO! Needless to say, I do not agree with her conclusion, and would certainly take issue with her on the use of the name 'fulvic acid' for the soil compound.

One of Budinger's findings that I did find intriguing was the relatively high amounts of oxalic acid present in the ring soil. This was something I had missed completely. While the presence of oxalic acid in soil can be considered very unusual, in the context of this case there may be a logical explanation. Oxalic acid is a well known component of highly efficient chemiluminescent reactions and, accordingly, it may offer a further corroboration that a chemiluminescent substance was deposited by the UFO. This possibility is explained more fully towards the end of my scientific report to which the interested reader is referred (see Appendix).

Budinger's analysis and conclusions were also published in JUFOS, following which a complete report on the Delphos case was published by the UFO Coalition in 2002. The latter consisted of all the available data on Delphos including my and Budinger's analyses combined into one volume. This last report again concluded that the case remained unexplained in prosaic terms.

Initial approaches to Nature and JBIS journals

Since the time of that last publication the case had remained largely forgotten which puzzled me since this was a rather unique case in which a proper analysis had been done and independently corroborated and which strongly suggested that it was genuine, even pointing to an unusual technology employed by the observed UFO. I was therefore confronted with a situation that I felt was totally unsatisfactory. So far the only people who were aware of the event were UFO enthusiasts and they had done little to promote its importance to the outside world. And in my opinion the people who needed to know about my findings were scientists who would be able to propose viable alternative explanations to the event, if indeed there were any. For that reason I decided that the best solution to this predicament was to get an updated paper published in a peer reviewed scientific journal. In June 2012 I approached the prestigious journal Nature and sent them copies of my JUFOS paper and original Ted Phillips case report accompanied with an email giving relevant background information. I ended the email with the following passage:

However, in view of the passage of the many years since the original event, it is ironic and puzzling that the case is still largely unheard of by the scientific community at large. I therefore submit these two reports to explore Nature's editorial interest in the case, and to determine whether you might deem it worthwhile to discuss its merits in your esteemed journal.

After some time had elapsed with me sending a request for a reply I received the following answer on the 20th signed Manuscript Submission:

Thank you for your email, and we apologize for the delay in response. In regards to your query, unfortunately Nature does not accept papers previously published elsewhere. For more information please view our duplicate submission policy at (ref. URL).

I replied on June 24th with the following:

I appreciate what you say about my paper having been published elsewhere. I have taken the trouble to check with the J.Allen Hynek Center for UFO Studies in the USA, and they have informed me that no more than 500 copies of the original JUFOS were distributed. This means that JUFOS had an extremely limited reach to the public and, furthermore, JUFOS is not indexed in the Web of Science (formerly the Science Citation Index of the Institute of Scientific Information). This means that JUFOS is, by definition, a low impact, obscure journal. In view of this, if anyone should now want to access my original report they are going to have great difficulty in doing so. I have also noted that in one of your web links you provide the following comment:

*"Consideration by the Nature journal is possible if the main result, conclusion, or implications are not apparent from the other work, or if there are **other factors**, for example if the other work is published in a language other than English.*

In view of the established obscurity of JUFOS, wouldn't it therefore be correct to place its original use in the 'other factors' category - equivalent indeed - to a work published in a different language.

Additionally, to make my paper more easily publishable in Nature I propose to amend it significantly at the beginning to properly introduce the subject matter to the Nature readership.

Because of the major implications presented in the paper it's likely to generate considerable interest in the scientific community at large, and for that reason alone its publication by Nature is warranted in my humble opinion.

I look forward to your reply.

I received the following on the 27th:

Thank you for your email. Following your comments, your article was passed to our editors, who are regretfully unable to offer to publish.

It is Nature's policy to decline a substantial proportion of manuscripts without sending them to referees, so that they may be sent elsewhere without delay. Decisions of this kind are made by the editorial staff when it appears that papers are unlikely to succeed in the competition for limited space.

In the present case, while your findings may well prove stimulating to others thinking about such questions, we regret that we are unable to conclude that the work provides the sort of firm advance in general understanding that would warrant publication in Nature. We therefore feel that the paper would find a more suitable outlet in a specialist journal.

We are sorry that we cannot respond more positively on this occasion, but hope that you will rapidly receive a more favourable response elsewhere.

Kind regards, Nature Administration

I sent them a further reply querying why they were "unable to conclude that the work provides the sort of firm advance in general understanding that would warrant publication" but it was obvious by now that I wasn't going to succeed using Nature as the journal. I therefore took their advice and approached a 'specialist journal' in the shape of the Journal of the British Interplanetary Society (JBIS). In fact, this had already published an excellent paper on UFOs in 2005 entitled *Inflation-Theory Implications for Extraterrestrial Visitation* authored by J. Deardorff, B. Haisch, B. Maccabbee and H.E. Puthoff. (see link below). I sent the editor a similar email and attachments to the one I had sent Nature.

http://www.ufoskeptic.org/JBIS.pdf

The reply I received was the following:

I want to thank you for your submission to JBIS and to congratulate you on your passion and interest in this subject. Unfortunately, the subject matter is not within the publication scope of the Journal of the British Interplanetary Society and so we must decline the papers. I wish you the best of luck in your future research.

I replied by referring to the Inflation Theory paper already previously published by JBIS. The editor responded that this had been done under a previous editor and it was not his intention to publish any further papers on UFOs:

Different Editors of the journal will have their own policy on the publication of papers on a specific topic. As I am Editor of JBIS, it is not my policy to promote the publication of UFO report papers, detailing specific events, sightings or related phenomena within JBIS. This is for two reasons:

- *(1) There are far too many claimed reports, for whatever reason, and JBIS is not the place to log such reports which will swamp the academic literature for the astronautical community that the journal aims to serve. There are UFO journals which serve this purpose as you are well aware.*

- *(2) Although I am very interested personally in the question of life, intelligent life, in the universe, I am highly skeptical as to the methods, techniques and standards of 'evidence' being justified as scientific, for the majority of UFO cases. This is not to say it isn't a worthwhile "phenomenon" to study, I just don't believe JBIS should be the home of such studies, where a higher standard of scientific rigour is required.*

I have not fully read your paper, because I simply don't have time to. I am already dealing with around 200 papers currently, on my own, so I don't have time to afford the luxury of reading your paper, although I am sure it is a good read. In addition, your paper was published in another journal, so why would you want to re-submit another version of it to a different journal. Please don't get me wrong, JBIS will publish SETI papers, which attempt to arguing the existence or not of intelligent life in the universe, but from a philosophical perspective and with a view to the wider implications.

I thank you for your interest in the journal, but I regret that we cannot proceed with your paper.

My somewhat lengthy reply has been shortened for clarity:

Thank you for your reply. Your comments betray a negative bias towards the subject of UFOs which is fairly common and can largely be attributed to misinformation or simple lack of knowledge on the topic. Fundamentally, there is a tendency to relegate it into one of a mainly psychological or hallucinatory manifestation - which is patently discovered NOT to be the case by anyone who bothers to study the phenomenon carefully. One of the major reasons for this mistreatment was the early involvement of the US Air Force in UFO investigation which knew it had a hot potato on its hands and desperately wanted to downplay interest by creating an atmosphere of ridicule.

You begin by stating that 'there are far too many claimed reports' (relative to what, may I ask?) and then imply that the JBIS might somehow end up being a repository of such reports! On the contrary, I am requesting that only one very special scientific report is published which describes the analysis and its conclusions of a ring of soil which has been grossly affected by a UFO hovering just above it. Because of the report's analytical nature your comment about 'a higher standard of scientific rigour' being required for the JBIS is fully satisfied. I am certainly not suggesting that the JBIS becomes a regular disseminator of standard visual UFO reports.

If you don't have the time to read my paper fully, then at least read the discussion section (i.e. the last five pages). The reason the report needs to be published in the JBIS is that it will enable it to get a much wider scientific readership and thereby increase much-needed searching debate on the subject.*

I did not receive a reply to this email, so I sent a reminder:

I would be grateful for your considered reply to my email sent earlier and reproduced below. In the meantime I have noticed that you have another paper published (under the heading 'Extra-terrestrial Studies') in the JBIS on a UFO theme regarding 'alien abduction' in the following issue: JBIS Vol. 68, No8, 2010 Is "Alien Abduction" Extra-terrestrial Visitation? Developing Prospective Study Designs to Gather Physical Evidence of Alleged "Alien Abduction" by Martin Hensher.

Again, I received no reply, so I sent another reminder:

I would appreciate the courtesy of a reply to my two earlier emails.

*Whether you like it or not there **is** a phenomenon out there which deserves proper scientific study, and my proposed paper would be a vital stimulus to initiate that study for which the readers of the JBIS would be hugely grateful.*

This is the reply I received:

I gave you my answer in the two previous emails I have sent you and I don't intend to enter into further dialogue on this matter.
I wish you the best with finding an alternative publication for your paper, but JBIS are not interested in publishing this specific one. I thank you for your interest.

To which I sent this response:

With respect, you did NOT give an appropriate answer - you instead decided to refer to UFO reports in general. My report is NOT one of those!

*My report is a **scientific study of a physical residue left by a UFO** - it's a whole new ball game which by implication deserves publication in a scientific journal.*

*Can you specifically explain why my (scientific) report as it stands cannot be published in the JBIS? Give me a **rational** reason which I can gladly consider.*

I did not receive a further reply at this point.

My challenge to SETI champion Seth Shostak

Then in September while browsing the internet I came across a podcast featuring Seth Shostak who is head of the SETI effort to contact extra-terrestrials in the U.S. In this he stated how he was keen to be shown just one very good physical evidence-based case for UFOs:

http://podcastufo.com/wp-content/uploads/2012/09/sat-22-Seth-Shostak-SETI.mp3

I took this up as a challenge and sent him an email introducing myself and providing Phillips' original Delphos report and my JUFOS paper.

This was his reply:

Thanks so much for sending the articles. I'm not a chemist, so can't really speak to how unusual this ring was ... I wonder if, given that this was in an agricultural area, you shouldn't also consider the possibility of contamination by one of the many chemical agents used in farming. And if someone were pouring into a bucket on the ground, for instance, generating some spillage, you'd get a ring.

But as I say, I'm not a chemist. And beyond that, the SETI Institute doesn't investigate UFO sightings (we don't have the staff ... we're a very small group). But I did find this interesting, and I appreciate the thoroughness of your description and analysis. That's quite unusual in this field, I think.

This was my response:

Thanks for your reply. The notion of the ring having been caused through spillage or the like was indeed foremost on my mind when I originally requested the soil samples from the investigator Ted Phillips. I thought that since I was an organic chemist it would be relatively easy for me to pin down any sort of mundane contamination from the types of materials you'd usually find in a farming environment. But this possibility was found absolutely not to be the case! The ring soil was instead discovered to be uniformly impregnated with a water soluble organic compound the main component of which (as identified through thin layer chromatography) was an air-sensitive coloured compound which rapidly oxidized to give a fluorescent product. These latter characteristics put the main component firmly into one of a special category – i.e. that of a potentially chemiluminescent substance. From this simple observation one could start to rationalize the sighting report.

I then added a summary rationale and emailed back to him.

He replied with the following:

Well, I'm about to go on travel, so no time for e-mail. But I think that -- aside from the speculation -- the only thing that will get the attention of many scientists here is a "smoking gun". Some aspect of the evidence that points strongly to extra-terrestrial origin. Odd arrangements, unusual materials -- all of those could have prosaic explanations, so won't be very convincing. You need to be able to point to one thing that's a clincher.

I replied with this:

The only thing I can think of as a 'clincher' is to do an isotopic analysis of the precipitated soil compound and check for anomalies. There may still be ring soil material available to do this.

Apart from that, the fact that the soil analysis corroborates so fully with what the witnesses claim to have seen is - in my opinion - worthy of much fuller discussion by the scientific community. Everything points to this case being a truly anomalous event. How - for example - does one explain the ring being skewed towards the wind direction on the night in question?? Simply brushing these considerations aside because there 'might' be a prosaic explanation is not good enough, because there are simply too many unusual factors that have to be taken into consideration and explained adequately.

I didn't get a reply to this email. But four months later I sent him another email to restart the discussion:

I was just looking over our email exchange back in September (see below), and you asked about providing 'a clincher' for extra-terrestrial origin. I suggested to do an isotopic analysis on the isolated soil material and wondered whether you would agree that this could settle the matter as far as the Delphos case was concerned (a multiple witness case with soil effects). If it would, why not show more interest and be proactive in investigating such unusual reports? I have attached for your consideration a photograph of the **actual glowing ring** *taken at night by Erma Johnson soon after it was deposited by the alleged departing UFO. The commentary below the photograph describes the circumstances of how the photo was taken. Apparently the ring glow was so bright that one could have read a newspaper by it.*

I find it remarkable that you show little **natural scientific curiosity** *to this subject. You are clearly convinced in your own mind that there's nothing to it, otherwise "thousands of scientists would be investigating it". But there is a chicken and egg situation that needs to be pointed out here! Scientists will be reluctant to start the ball rolling due to basic fear of their peers' negative reaction, which in turn arises because of the aura of* ridicule that's associated with the topic. Are you aware that the US Air Force had a major influence in instilling this ridicule in the first place?

He now replied:

What I would suggest is that you submit this to a refereed journal. At that point you'll get some action from scientists. I get stuff like this every day -- there's no way I can investigate it all. And I must also state that it's unclear to me why glowing lights or radioactivity are automatically assumed to be evidence of alien craft.

My response:

I had in fact already approached Nature journal some months ago to see whether they would be interested in publishing an updated version of my soil analysis report. The response I got was that it would be more appropriate to publish in a specialist publication, so Nature couldn't oblige in this instance. I then approached the editor of the JBIS, this being what I imagined to be an appropriate 'specialist' journal. Indeed, it had already published the occasional paper on the UFO topic in the past.

Unfortunately the current editor of the JBIS is a young guy who was clearly more concerned about his own credibility - and flatly refused to get involved with publishing papers on UFOs, period! So we come to an impasse – this subject is treated with scorn, with the result that any attempt to bring important cases to scientific scrutiny doesn't get past the first post! And then scientists such as yourself are justifiably able to proclaim that 'there's no good evidence' for UFOs!! Do you see the dilemma here!?

Incidentally, when you state that "I get stuff like this every day" are you referring to cases where UFOs have actually left physical marks or residues in the ground, or simply referring to 'lights in the sky' type reports on which nothing much can be done?

With regard to your query as to "why glowing lights or

radioactivity are automatically assumed to be evidence of alien craft" - I don't believe any responsible investigator would make such a sweeping claim. Each UFO report would have to be judged on its merits – some are very poor indeed, being little more than misidentifications of known aircraft and/or aerial phenomena. It's the 5-10% of close encounters with unknown structured craft - in many cases seen by very credible or multiple witnesses - that are the core problem with this subject. Add to that the numerous instances of radar + visual cases and almost 5000 cases worldwide where alleged UFOs have left imprints/marks on the ground and you start seeing the scale of this problem. The 'glowing' issue arises because many UFOs happen to be brightly illuminated, so it makes sense to study the possible causes for the light emission. I consider the Delphos case that I personally investigated through chemical analysis to sit at the very top of this core category of high strangeness reports.

His reply:

Too bad, Erol. JBIS would have been one of those I would have recommended. You could, of course, ask for recommended journals from MUFON -- don't know whether they ever publish anything. I know that it sounds as if the world is against this, but that's really not true. The International Journal of Astrobiology publishes stuff that's controversial if the referee feels that the evidence presented is decent. Witness testimony is generally useless, by the way. Highly unreliable.

My response:

I will approach the JBIS editor again and ask whether he might consider changing his mind. If you have no objection I will tell him that you yourself considered JBIS to be an appropriate refereed journal for an updated version of my report and see what his response is.

I must admit to smiling at your denial that the 'world is against

UFOs' comment. Indeed, it is patently obvious that the scientific world IS against it, as evidenced from the (largely) sniggering and condescending attitudes that most (uninformed) persons in the scientific community display towards the subject. The early introduction by the US Air Force of the ridicule element when referring to UFO reports has a great deal to answer for!

My own view of witness testimony is that it simply comes in varying degrees of credibility and worth. Obviously a single witness who's just come out of a bar in an inebriated state is not likely to score highly in that regard! On the other hand a group of police officers who report seeing the same thing from different vantage points will justifiably be at the other end of the scale, and needs to be taken more seriously! Many UFO reports have come from experienced pilots and military people whose observational prowess is likely to be much higher than that of a lay person. Even the famous astronomer Clyde Tombaugh had a sighting: (Ref URL)

The notion that ALL witness testimony is unreliable is therefore patently untrue, but is falsely used by debunkers to justify the summary dismissal of this subject and perpetuate its taboo status. It's also an outrageous slur on those people who have seen something very unusual and have risked derision to report it. This is NOT the way scientific investigation should be working! If hundreds of people claim to have seen a very large boomerang object with lights on its 'wings' fly over Phoenix, Arizona and their descriptions tally even though seen from completely different points of the city – such as happened on March 13, 1997 – the scientific community should be perking up its ears and taking note! Here's an interesting link – check out the first 15 minutes or so of the following documentary which features testimony of the Phoenix boomerang sightings: (Ref URL)

To summarize, any SINGLE witness statement can be glibly dismissed, but multiple witness testimony from high calibre witnesses is a whole different ballgame and should be considered far more seriously in my view.

I will contact the JBIS editor again regarding the publishing of my Delphos report and let you know the outcome.

A second attempt to publish in the JBIS

So in February 2013, eight months after I had originally approached him, I sent another email to the Editor of JBIS:

I thought you'd be interested to know that I have been in touch with Seth Shostak regarding the Delphos case that we ourselves had discussed some months ago. He has read my report, and has also seen a photograph of the glowing ring soil that was taken soon after the UFO had departed from the area. I have attached this photo for you to also view, since it's an important part of the evidence supporting the sighting report.

Seth's view was - unsurprisingly - that this information should be published in a refereed journal so that it could be considered by other scientists ......So if Seth Shostak also believes that the JBIS would be an appropriate journal to publish an updated version of my report, does this make it more relevant to consider my request for publication in the JBIS?

His reply:

I appreciate your persistence and belief in your own research. I'm afraid it's not a matter of the quality of your paper and any recommendations from highly established researchers such as you illustrate. So please don't take the rejection personally. It is simply my editorial policy to not publish UFO papers. This is not because I don't think that field of study is worthy but because I think there are other more fitting journals which specialize in this area and JBIS does not, by my choice.

When I asked what 'more fitting' journals there were he suggested UFO centric ones such as the European Journal of UFO and Abduction Studies. But I again reminded him that the whole point of this exercise was to get the paper published in a main stream scientific journal so that a much wider readership could be achieved encompassing those who were not already familiar with the topic.

His reply came back:

I sympathize with your situation but my decision stands. I took the paper to a wider consultancy among some of my advisory board and they all advised it was not for JBIS.

Understand my problem too which is actually similar. I'm an advocate of interstellar studies and that subject is also sometimes considered on the fringe by the main stream, as indeed is SETI. So I have to work hard to ensure rigour in the papers. If I was to allow your paper then it makes my goal here even harder.

Please do not take this as a comment on your paper or its quality. It has not been reviewed by me or others. It is just my editorial policy to not publish UFO papers. And if I did, I can assure you I would receive many letters from readers complaining and even many people resigning from the BIS in protest. I can't take such risks at this time.

My reply:

OK - fair enough. It's a curious state of affairs that a topic considered early on by a senior atmospherics physicist (James E. McDonald) to be 'the greatest scientific problem of our times' might actually lead to resignations from the BIS on its mere mention in your journal. Clearly there is one hell of a mountain yet to climb to get the respectability that this topic deserves!

His reply:

I agree it's one of the biggest scientific problems. Keep in mind that NASA and ESA scientists publish in JBIS. How would they feel about it? It's a tough one. Need to maintain academic credibility. But I say again, this doesn't devalue the worth of your work. Keep doing it.

I thanked him and signed off, thinking: I'll 'keep doing it' to but what end?

An approach to publish in the International Journal of Astrobiology

Since JBIS wasn't possible, I wondered which other journal I could approach for publication. Seth Shostak had mentioned the International Journal of Astrobiology (IJA) and indicated that its editor wasn't averse to publishing controversial papers as long as they were well evidenced. I resolved to approach this journal. But first I had to update and revise my original Delphos report, which I did by writing a new introduction and adding additional information, mainly pertaining to Phyllis Budinger's analysis. This is the version which can be found in the Appendix at the end of this book. I created an account at the IJA Manuscript website in February 2013 and submitted my paper at the very end of May.

I received the following response on the 5th June 2013:

I am sorry to inform you that your manuscript # IJA-AR-13-0439 entitled "The search for signs of extraterrestrial intelligence on earth: strong chemical and physical evidence for the existence of an unconventional luminescent aircraft (commonly called a UFO) observed by multiple witnesses at a farm in Delphos, Kansas, USA." has not been accepted for publication in International Journal of Astrobiology.

This is not a rigorous research paper. It lacks data and hard experimental evidence. The references are in the wrong format and cannot be found in the peer reviewed literature. Thank you for considering the International Journal of Astrobiology for the publication of your research. I hope the outcome of this specific submission will not discourage you from the submission of future manuscripts.

Somewhat shocked at this I replied thus:

Thank you for your reply. When you say it lacks data and hard experimental evidence, can you please explain what is wrong with the data that I have presented on the soil? Also the additional information such as the photographs of the ring soil and its dimensions, particularly the way that the ring shape is skewed towards the wind direction on the night of the event.

I look forward to your reply.

I did not receive a reply to this, so sent a further email on the 17th:

Further to my response of the 6th June (see below) could you please explain in specific terms what is wrong with the data that I have presented in my paper. You will no doubt appreciate that I have expended a lot of effort to put this manuscript together. Furthermore it relates to a topic that has so far been largely ignored by the scientific community (due to lack of data!) and it therefore merits special consideration for publication, and your journal is probably the most appropriate for this (also suggested to me by Seth Shostak.

Could you therefore please let me know what specifically are the data issues with the publication of my paper.

This is the reply I received:

The conclusions you reach require extraordinary care in the analyses. Details of the analyses are not presented as well as standards and control experiments. Indeed, if true this would be an extraordinary finding, and as such would also require independent verification.

My response:

*The independent verification **has been done** by Phyllis Budinger as detailed in my report. The fact remains that the analysis data can be used to formulate a model that **fits** with the sighting report in **all** its salient features, and also explains the ring's characteristics - **including** the fact the ring is skewed towards the wind direction* on the night of the event!! How else can the ring be explained? How would you try to explain it?

His reply:

What I meant was that we need detailed chemical analyses not just morphological analyses. In my opinion you have the ground for the formation of a hypothesis, but not the proof. Suggest a few alternative hypotheses and test them experimentally as well.

My response:

*There is enough data presented to pinpoint another explanation if there was one. The fact that the isolated compound oxidizes readily and gives rise to a fluorescent product puts it into a **special category**, i.e. a potentially chemiluminescent compound, and because of that it's not the sort of material that could be readily purchased and dumped into the ground by anyone!*

*If there is someone out there who could suggest another explanation, I would be interested to hear it. But in order to do that, the paper needs to be published to allow alternative explanations to be forwarded. I cannot myself - **in all honesty** - suggest an alternative hypothesis!*

He didn't reply, so I sent him a reminder six days later:

You didn't reply to my last email (shown below) and I naturally still wish to know whether there is any chance of my paper being published by your journal. The information the paper contains needs to be made aware to other scientists who have an innate interest in the possibility of life elsewhere in the universe, and your journal is therefore the most appropriate for the paper's publication.

His immediate reply:

As your paper stands now it is not publishable in the International Journal of Astrobiology. You might try Astrobiology published by Mary Liebert.

I thought, here we go again - OK - let's try Astrobiology instead.

An approach to publish in Astrobiology journal

I went through the same submission on their website creating an account on June 24th and submitting on the same day.

I received a response on July 10th 2013:

I write you in regards to manuscript # AST-2013-1053 entitled "The search for signs of extraterrestrial intelligence on earth: strong chemical and physical evidence for the existence of an unconventional luminescent aircraft (commonly called a UFO) observed by multiple witnesses at a farm in Delphos, Kansas, USA," which you submitted to Astrobiology.

We are unable to publish your manuscript in Astrobiology as UFO's do not fall within the scope of our Journal. Your manuscript belongs in a more specialized journal.

Thank you for your interest in Astrobiology.

My immediate reply:

To those scientists that have taken an unprejudiced and detailed look at the UFO phenomenon the possibility that this represents our planet being covertly monitored by ET is a highly regarded one. What has been missing in order for mainstream science to actively consider this notion has been the lack of good physical evidence, which my report now hopefully provides. Are you proposing that evidence for manifestations of ET on our own planet DOES NOT come under your Journal's stated theme: "...to advance our understanding of life's origin, evolution, and distribution in the universe"!

I note your suggestion that my report belongs to a more specialized journal. I would be interested to learn what you would propose for this. If you are hinting at a UFO centric journal the whole point of this exercise is to get mainstream science to be informed and enable the majority of uninformed scientists to open their eyes to the evidence, and this will only happen if a journal such as yours publishes the material, rather than publishing in one with a necessarily limited readership.

It is high time that this subject be treated in a proper and dignified scientific manner and I would respectfully ask that you reconsider your refusal to publish my report. You state in your email that UFOs do not fall within the scope of your Journal, but if they do represent the presence of ET on earth then they most definitely should be, don't you think?

The editor did not reply to this, and I didn't pursue it.

A second attempt to publish in Nature journal

Since I had updated my report I now made a second attempt to get it published in Nature by formally applying through its website, which I hadn't done first time round. I got back a reply from a senior editor the following day:

Thank you for submitting your manuscript entitled "Strong chemical and physical evidence for the existence of a luminescent aircraft observed by multiple witnesses at a farm in Delphos, Kansas, USA for consideration. I regret that we are unable to publish it in Nature.

As you may know, we decline a substantial proportion of manuscripts without sending them to referees, so that they may be sent elsewhere without delay. In such cases, even if referees were to certify the manuscript as technically correct, we do not believe that it represents a development of sufficient scientific impact to warrant publication in Nature. These editorial judgements are based on such considerations as the degree of advance provided, the breadth of potential interest to researchers and timeliness. In this case, we do not feel that your paper has matched our criteria for further consideration. We therefore feel that the paper would find a more suitable outlet in another journal.

My response:

Since this manuscript potentially implicates the presence of an advanced technology from an unknown source I would be very grateful if you can clearly explain why you have decided it "<u>doesn't represent a development of sufficient scientific impact to warrant publication in Nature</u>"!!!?

Since the search for ET 'out there' is among the foremost scientific research programmes in the world today why is it that the notion that they may already be here covertly monitoring human civilization receiving such scant regard? The only way that science can grapple with this topic is for its adherents to <u>become aware of what's going on</u>, in the way of significant evidence, which my manuscript provides.

His reply:

Thank you for your letter asking us to reconsider our decision on your manuscript entitled "Strong chemical and physical evidence for the existence of a luminescent aircraft observed by multiple witnesses at a farm in Delphos, Kansas, USA.". Now that I have had a chance to discuss the matter carefully with my colleagues, I am sorry to have to tell you that we cannot reverse our original decision.

Unfortunately, the number of papers with possible claims on space in Nature vastly exceeds the number that we can publish each week, and we are therefore frequently forced to make difficult decisions. Please see (Ref URL) for a general explanation of Nature's editorial procedure.

I am sorry that on this occasion we cannot be more positive.

My response:

Thank you once again for your reply. However, you do not focus on the matter at hand. This is not a 'normal' report - it provides evidence for the existence of an aerial device that shouldn't be there! It has long been claimed that 'there isn't any good evidence for UFOs' - hence the reason why scientists in general do not recognize the reality of the phenomenon. Well, here is the evidence!! - which I would imagine would be of great importance to the vast majority of scientists that read your journal. Why do you not recognize the uniqueness and huge significance of the findings described in my report?

Please refer specifically to the report's content and its significance rather than classifying it as 'just another paper' which you don't have space for - as you infer in your reply.

He didn't reply to this, and I therefore wrote an email to the Editor-in-Chief of Nature to protest at this editor's indifference, but he never replied either.

It was very obvious by now that I would not succeed in getting my paper published in a main stream science journal, and after consulting with Rodeghier and Clark in the U.S. I decided to get my results out to the wider public in the form of a book. I hoped that this might generate sufficient interest to force the scientific community to consider the merits of the case as well as of UFOs in general. As a final postscript to this debacle, two months after the rejection by Astrobiology I received notification by email of a paper just published by that journal. I took umbrage at the title and complained to the editor:

I have just received notice of a recent paper published in your journal entitled: The Quest for Extraterrestrial Life: What About the Viruses?

The fact that you can publish a wholly speculative paper such as the above while wilfully ignoring actual strong evidence for the physical existence of a 70 year old world-wide phenomenon known as UFOs is beyond belief! I am now being forced to publish my results in the form of a book, part of which will be devoted to the blindness and sheer prejudice of Big Science towards this important subject. Are you aware that a senior U.S. atmospherics professor (James MacDonald) who chose to investigate UFOs concluded by calling it 'the greatest scientific problem of our time'!! Are you also aware that the governments of France and Chile have been actively investigating UFOs and recently announced that they will be pooling their resources to achieve their common aims.

Meanwhile journals such as yours refuse to treat UFOs seriously with the result that scientists can continue to falsely believe that there's nothing of substance to it!

This is a pathetic state of affairs which urgently needs to be addressed.

Needless to say I didn't receive a reply to this either!

My concluding communication with Seth Shostak

I decided to write to Shostak and inform him of my unsuccessful attempts to get my paper published.

His reply on the 26th August 2013:

Thanks for the e-mail. I admit (and wrote ...) that it might be hard to get UFO papers in mainstream science journals. That's not because they are "inadmissible" ... that's not the way the editors think. But they do have a high standard for work that makes "extraordinary claims", and that's as it should be. Physical evidence will trump witness testimony every time -- indeed, I've been on panels with some of the people you cite below, and can tell you that they are definitely not convincing to anyone with a science background. They've seen something – but that's not the point.

You could go for a self-published book. I have a stack of those, actually -- so yours wouldn't necessarily stand out, unless what you say is quite different. Better to find a journal ... really, it is. You can't give up just because of two rejection slips. J.K. Rowling sure didn't.

My reply:

But my report actually DEALS with physical and chemical evidence!!! In fact, excellent evidence that actually explains the ring soil and its elongation towards the wind direction on the night of the event.

Can you therefore please offer an explanation as to why my physical evidence-based report has been refused publication by FOUR science journals (not two), apart from the dogma that they will simply NOT DEAL WITH anything concerning UFOs!!

His response:

You'd have to ask the editors, Erol. I can't speak for them. It's too easy to say that this is "dogma". In my experience, that's almost never the correct answer...

It was clear by now that all this correspondence with bona fide scientists was leading nowhere. I reminded myself why I had approached Shostak in the first place. He had requested in his podcast to be sent information on just one very good physical evidence-based case for UFOs. So I sent him a final email with the following query:

OK – here's my challenge to you:

You have my full report detailing the chemistry of the ring soil as well as the photographs of the glowing ring and the ring itself taken by the sheriff the next day. The latter shows an elongated white ring approx. 8ft diameter and skewed towards the wind direction on the previous day.

You have all the data in front of you. What is your suggestion for the ring's cause?

I look forward to your considered opinion.

As I suspected, he didn't reply to this. He had made a noble request for someone to send him information on just one excellent UFO physical trace case. I sent him that information, and he ignored it!

The fact that science continues to pointedly disregard the UFO phenomenon is nothing short of scandalous in my opinion and my great wish is that this book will spur those open-minded enough to initiate a serious debate on what could eventually lead to a major advance for human kind.

The challenge that I posed to Seth Shostak is now directed to anyone else out there who might care to offer a possible solution…

APPENDIX

The following paper was submitted for publication in Nature, Journal of the British Interplanetary Science (JBIS), The International Journal of Astrobiology, and Astrobiology journal but was turned down because the subject matter was deemed 'inappropriate'.

The search for signs of extraterrestrial intelligence on earth: strong chemical and physical evidence for the existence of an unconventional luminescent aircraft (commonly called a UFO) observed by multiple witnesses at a farm in Delphos, Kansas, USA.

EROL A. FARUK PH.D

Abstract: The on-going controversy over the origin and nature of many thousands of UFO reports submitted by persons from all walks of life around the globe for the past 70 years has been compounded by the utter failure of anyone so far to come forward with incontrovertible physical evidence for the existence of such craft. This paper describes a multiple witness observation of a UFO together with the chemical analysis of an affected ring of soil that the UFO allegedly left behind which was profoundly altered in its chemical and physical characteristics. The results from this analysis strongly corroborate the sighting report in all its salient features and suggest the presence of a hitherto unknown technology.

INTRODUCTION

It is generally accepted that the modern UFO reporting era commenced with the observation of unknown aerial phenomena apparently under intelligent control during World War II when allied bomber pilots found themselves being closely followed by glowing balls of light, seemingly monitoring their activities. The pilots adopted the name 'foo-fighters' for these phenomena. After the war sightings began to be reported of metallic disc shaped aircraft from around the world, which peaked in the mass sightings over Washington DC in 1952. These latter observations were accompanied by detection through radar, leading one senior air-traffic controller at the city's airport to state that: "their movements were completely radical compared to those of ordinary aircraft". The events inevitably made front-page headlines in newspapers around the country:

http://en.wikipedia.org/wiki/1952_Washington,_D.C._UFO_incident

Since that early phase the UFO phenomenon has firmly established itself as a modern day mystery. Sightings continue to occur across the world and these are notable for their consistency both in the appearance and perplexing flight characteristics of the objects. Typically they are described as disc, cigar, spherical or triangular in shape, and either metallic or luminous in appearance while also silent or emitting a slight humming noise. In flight, they can be slow moving or stationary and then accelerate and depart in a matter of seconds. One very notable characteristic often observed is sharp right-angled turns while in flight, seemingly oblivious to the severe inertial stresses that would result. When confronted with these reports the normal instinct would be to dismiss them as observational aberrations, but the fact remains that in many instances the same object or objects may have been reported by multiple witnesses from different vantage points. A classic example is the Belgian wave of 1990 in which thousands reported seeing unusual triangular craft with lights, which were also tracked on radar and chased by air force F-16 jets.

http://en.wikipedia.org/wiki/Belgian_UFO_wave

In 1999 a ground breaking 90-page French report entitled: "UFOs and Defence: What must we be prepared for?" was published under the heading COMETA following a high-level investigation on UFOs conducted by former senior officials of the Institute of Advanced Studies for National Defence (IHEDN). The authors consisted of senior military personnel and scientists and the report concluded that 5% of the cases they studied were inexplicable and that the best hypothesis to explain them was the extra-terrestrial one:

http://en.wikipedia.org/wiki/COMETA

http://www.cufos.org/cometa.html

One other notable feature of the UFO phenomenon is the large number of instances wherein the object has been observed close to or actually sitting on the ground, leading to potential physical evidence as a result. Ted Phillips in the U.S. is a leading investigator of such cases and has catalogued over 4000 to date. There are numerous examples of UFOs having left patches of scorched vegetation or earth or 'landing marks' from these types of reports, one of the most studied being the Trans-en-Provence case of 1981:

http://www.ufoevidence.org/cases/case110.htm

http://en.wikipedia.org/wiki/Trans-en-Provence_Case

Another type of effect from these 'close encounters' is vehicle interference in which an automobile engine will inexplicably stutter and die while in close proximity to a UFO. Both this and other types of physical evidence for UFOs were extensively reviewed in a scientific workshop chaired in 1997 by a leading astrophysicist Professor Peter Sturrock of Stanford University (Sturrock, 1997). The published Proceedings from this workshop can be viewed here:

www.scientificexploration.org/journal/jse_12_2_sturrock.pdf

Sturrock has also written a book based on these Proceedings (Sturrock, 2000)

Since physical evidence could potentially provide the required proof for the existence of UFOs it becomes necessary to seek out and investigate such evidence wherever possible, which is the reason this author took up the challenge in 1977.

BACKGROUND TO THE CASE

This author is a retired pharmaceutical development chemist who acquired an interest in UFOs during childhood while keen to learn about astronomy. This curiosity was heightened after reading about this particular case in FSR Case Histories (Phillips,1972a) and was sustained until, in 1977, my employment as a postdoctoral research chemist at Nottingham University led me to contact, via a UK intermediary, Mark Rodeghier of the Chicago based Center for UFO Studies (CUFOS). He, in turn, requested the principal MUFON investigator of the case Ted Phillips to send soil samples for me to analyze. On receipt of the material I advised my academic supervisor Dr. Barrie Bycroft (now a retired professor; contact details of whom can be supplied on request) of the case particulars and outlined the unusual characteristics of the affected soil. He became intrigued and referred me to the resident expert on chemiluminescence Dr. Frank Palmer (now deceased). After discussing with the latter the case particulars and odd nature of the soil, he also became keen to help analyze the latter, and was personally responsible for recording the fluorescence spectra presented in this report.

THE ORIGINAL CASE REPORT

This case is well known in the UFO literature having initially been reported in depth by the aforementioned Ted Phillips (Phillips 1972a, 1972b, 1981). For present purposes a brief description of the important aspects of the sighting report will suffice.

On November 2, 1971, at Delphos, Kansas, at 7:00 p.m. local time, a sixteen-year old boy, Ronald Johnson, was tending his sheep at his father's farm when he heard a rumbling noise and then stepped out of a small shed to see an illuminated object hovering beneath a tree about 75 ft. away from him. The object had an estimated diameter of 9 ft. and appeared to be about 10 ft. high (Fig. 1). The noise was likened to that of a washing machine that vibrates when out of balance. The boy described the object as multicoloured with blue, red and orange glows about its entire surface as it hovered 2-5 feet off the ground. He also observed a bright glow between the object and the ground, as though a shimmering material was falling off from the object. The boy said that it hurt his eyes when looking directly at the object and for several days after the incident his eyes were painful and he suffered headaches. After some minutes the object began to move off passing over a nearby shed by about 4ft at which point it emitted a high pitched sound like that of a jet aircraft (Fig. 2). Ronald briefly lost his vision while observing the illuminated object, but after his sight returned he looked into the sky and could see the bright circular light receding into the distance.

Ronald ran back to the farm and told his parents what he had seen. They didn't believe him at first and he became agitated. They all then went out to observe in the southern sky a bright luminescent object receding into the distance bearing "the colour of an arc-welder." It was estimated to be at least half the size of the full moon, which could also be seen in the south-eastern sky at the same time. Although this was the last time that the family members saw the object another witness Lester Ernsbarger, a reserve police officer in Minneapolis (ten miles south of Delphos), called in to his base to report a light in the sky which he had observed at 7.30pm in the northern sky from his position thus providing independent corroboration for the presence of the illuminated object (Fig. 3).

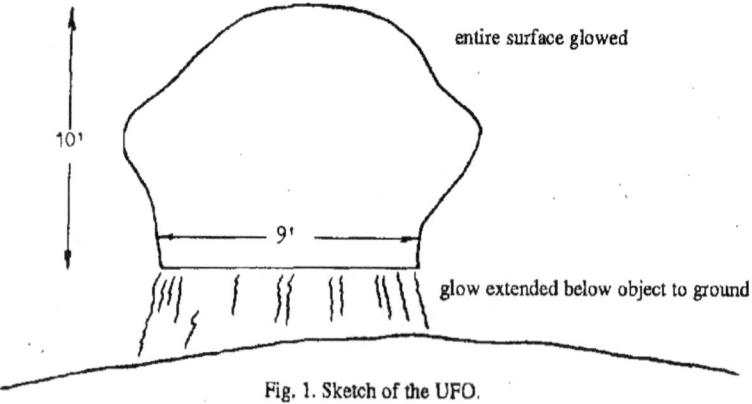

Fig. 1. Sketch of the UFO.

The Johnson family then proceeded to the site where the object had originally been observed. As they approached they saw a ring of soil over which the UFO had hovered glowing in the dark. Portions of the nearby trees reportedly also glowed. The glowing ring so impressed Mrs. Johnson that she ran back to the house to fetch a Polaroid 104 camera with which she took a photograph of the effect (Fig.4). No flash attachment was used, as the light from the ring was described as bright enough to read a newspaper by. She had only one exposure remaining in the camera and could see the illuminated ring through the view finder well enough to get the camera positioned. They claimed that the glow from the ring lasted long enough to be visible the following night.

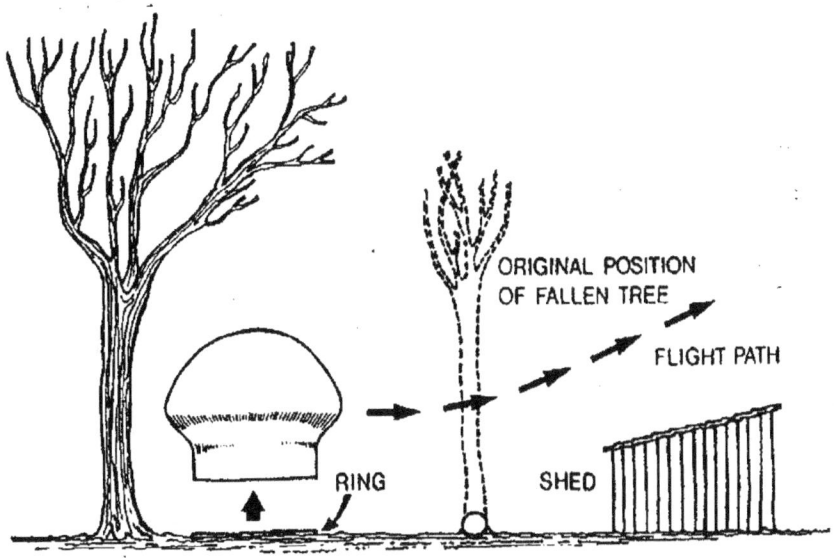

Fig. 2: Schematic of UFO departure.

The witnesses proceeded to touch the ring, which they described as having a cool, slick, crust-like texture, as if the soil was "crystallized", as well as being pitted with tiny craters or holes on the surface. They noticed an unfamiliar odour on their fingers and immediately experienced a numbing effect similar to that of a local anaesthetic which took several days or weeks to wear off. The Johnson family reported the alleged event to the local sheriff, who took samples and a photograph of the ring the next day (Fig. 5), and to a newspaper reporter.

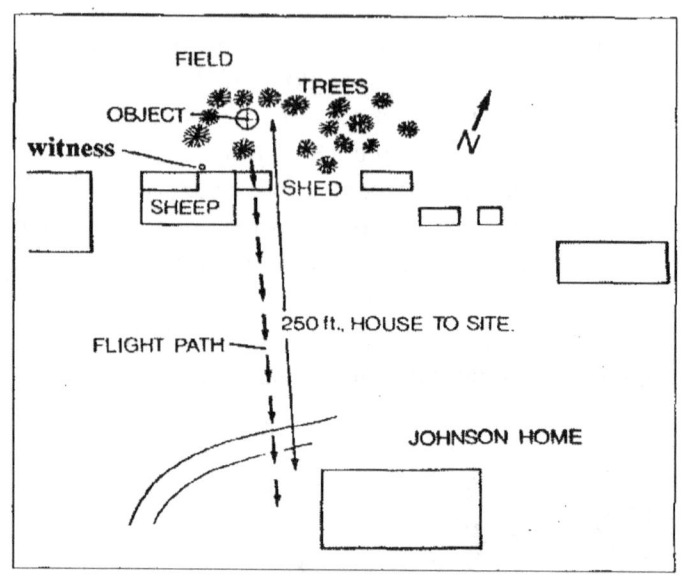

Fig.3: Site plan showing UFO and farm area

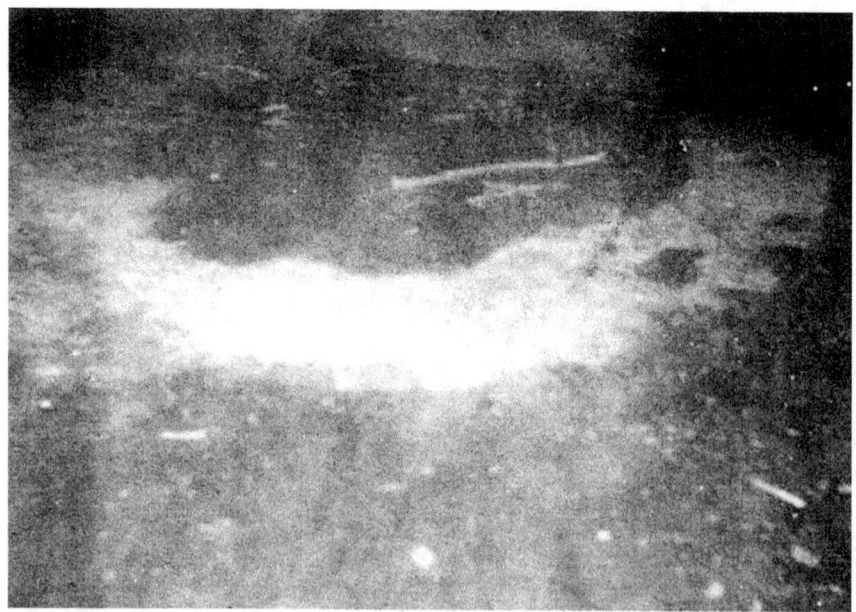

Fig. 4: Polaroid photograph of glowing ring (taken without a flash attachment)

Because of the high degree of evidential information inherent in the report it was considered by many to be a key case and samples of ring soil were duly sent to various laboratories in the U.S. to

Fig. 5: Photograph of ring taken 19 hours after the alleged event.

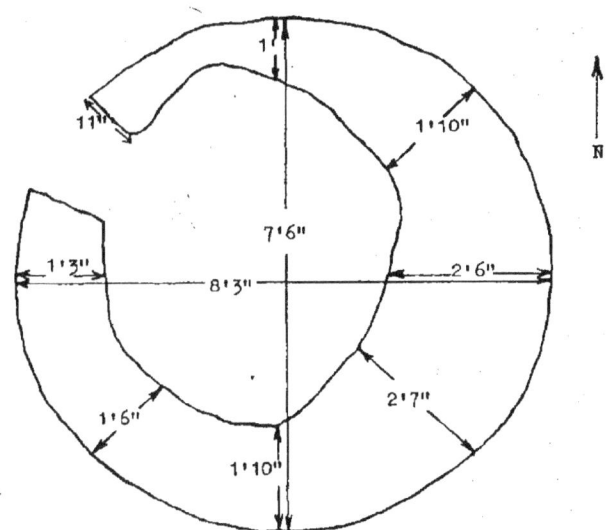

Fig. 6: Dimensions of ring soil.

analyze chemical changes that had arisen to cause the ring soil absorption anomalies. Unfortunately, because of lack of sufficient funding and resources none of these laboratories were able to offer any clues as to the nature of the luminescent nature of the ring soil this author suspected that an organic chemical explanation was likely, which was the motive for requesting the material to study.

MATERIALS AND METHODS

Numbered soil samples taken from various locations and depths of the ring one month after the alleged event were received from the principal investigator Ted Phillips in late 1977. A sample of surface control soil taken several feet outside the ring was also sent for comparison purposes. All samples had been kept refrigerated and sealed in airtight opaque containers to minimize decomposition. The manipulation and analysis of the soil was carried out employing standard organic chemical methodology.

RESULTS

(a) Ring Soil Hydrophobicity

Samples of ring soil were initially inspected for their alleged hydrophobicity. They did indeed appear to be surprisingly impervious to water, contrary to the usual expectations for soil matter. Any added water spontaneously formed into globules on the soil surface, similar to those that would form on a glass surface. Enforced agitation, however, induced the aqueous soil mixture to disperse with a resultant loss in the hydrophobicity of the soil. Centrifugation of the suspension separated out the soil matter leaving a clear brownish-yellow to red solution that was mildly alkaline and foamed on shaking such as a solution of household detergent would. These properties were consistent with the dissolved substance being an alkali metal salt of an organic acid and subsequent chemical manipulation supported this assessment. Such a compound would consist of a hydrophobic organic residue - R linked to a hydrophilic carboxylate anion $-CO_2^- M^+$ (where M represents an unknown metal ion):

$$R-CO_2^- M^+$$

hydrophobic   hydrophilic

This substance and its direct decomposition products appeared to be the only ones 'foreign' to the soil and were therefore solely responsible for the observed hydrophobicity. Extracting the ring soil with any organic solvent gave rise to a negligible amount of soluble material. A similar negative result was obtained when the control soil was extracted with water. The apparent paradox of a water-soluble – indeed highly soluble - substance being responsible for the hydrophobic nature of the ring soil can be explained in terms of a surfactant phenomenon. The hydrophilic sites normally present in the soil matter are bound up through non-permanent 'hydrogen bonding' with those of the substance enabling the organic residue R to behave as the new 'surface' of the soil (Fig. 7). The action of wetting the soil disrupts the hydrogen-bond linkages, allowing dissolution of the substance while exposing the soil hydrophilic sites to re-absorb water.

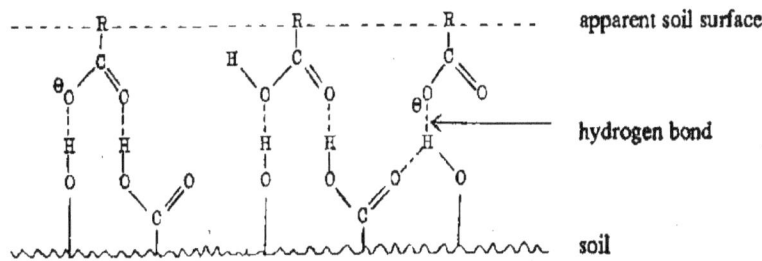

Fig. 7: Hydrophobicity of ring soil.

To test this idea further a number of organic acids were converted into their sodium salts and their aqueous solutions mixed with some of the control soil. After allowing the mixtures to dry water was placed on the surface of each sample to test for any acquired hydrophobicity. Compounds such as sodium benzoate (1) and sodium naphthoate (2) did not produce any appreciable effect, whereas sodium anthraquinone-1-sulphonate (3) imparted to the soil a certain degree of hydrophobic character. This indicated that the increased molecular size of (3) combined with the presence of additional polar carbonyl groups (a) able to participate in hydrogen-bonding and thus bind with the soil in 'template' fashion was causing the hydrophobicity. This finding becomes significant when the chemical nature of the ring soil compound itself is considered to explain the unusual degree of hydrophobicity apparent. Apart from producing the desired effect, the test also revealed how such a change in soil absorption could be brought about by the application of an aqueous solution of a suitable compound from an external source, an observation that may be significant.

(a) Chemical Manipulation of Ring Soil Extracts

To initiate the analysis of the unknown compound present in the ring soil extracts, the latter were each treated with a slight excess of an aqueous solution of silver nitrate. An immediate precipitation of the corresponding silver salt (4) of the presumed carboxylic acid occurred which was centrifuged off and washed, first with ethyl alcohol to remove water, then with diethyl ether and dried with a stream of nitrogen gas. The quantities of silver salt isolated in this way varied for different ring samples (Table 1).

Table 1. Silver Salt Precipitation

Soil sample	Quantity of silver salt precipitated (averaged): mg per gram of soil
*2	34.8
*3	12.8
*10	10.6
*7	10.1
Control soil	Negligible precipitation observed

Treatment of the silver salt with excess methyl iodide (MeI) in acetone for several hours afforded the corresponding methyl ester(5) which was soluble and imparted a yellow colour to the solution.

$$RCO_2^- M^+ + AgNO_3 \rightarrow RCO_2Ag\downarrow \xrightarrow{MeI} RCO_2Me$$
$$(4) \qquad\qquad (5)$$

Analysis of the solution was performed using thin layer chromatography, a technique that enables the physical separation of the components of a mixture in the form of 'bands' or regions of discrete compounds. Analysis in this way revealed one major yellow band of retention factor (Rf) of ca.0.5 (stationary phase: silica gel HF 254; mobile phase: methanol/acetone/benzene in proportion 20:5:25) situated among a continuum of bright blue fluorescent bands when the chromatogram was viewed under ultraviolet (366nm) light. This yellow band was very unstable and readily oxidized within minutes to give an almost colourless product. When viewed under 366nm light the product from this oxidation had identical fluorescence to the others, thus merging with them (Fig. 8). Covering the chromatogram with a glass slide to reduce exposure to oxygen slowed down this process considerably. The yellow band could be eluted off the chromatogram before it had time to fully oxidize and its UV-visible light absorption spectrum recorded in ethyl alcohol. This showed maxima at 220, 290 and 358nm (Fig. 9, Curve a). The two longer wavelength bands can be seen to decrease with time while there is a simultaneous increase in the 220nm band. After 45 minutes in the cell the absorption curve in the visible region has decreased considerably (Fig. 9, Curve b).

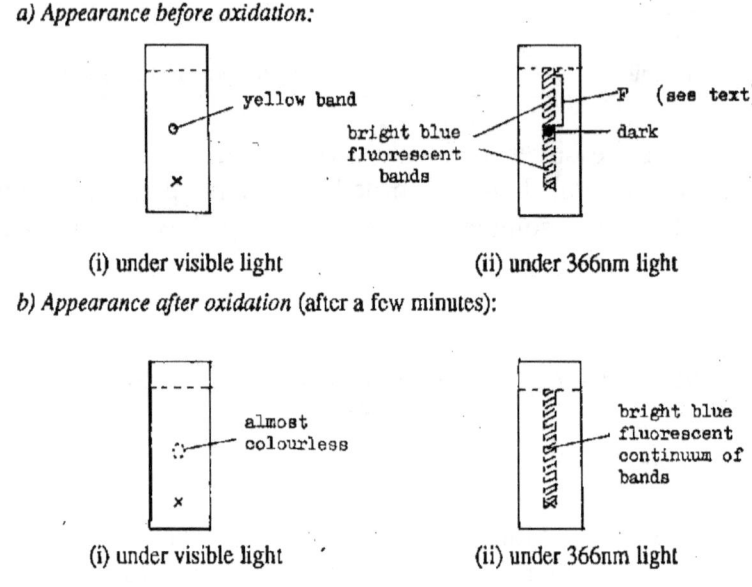

Fig. 8

This change represents a decolourization process and parallels that occurring on the chromatogram during oxidation. That oxygen is indeed implicated for the change will become more evident in due course. An assumption that can be made on basis of the observed spectral changes is that the 220nm band arises from the oxidation product whereas the 290 and 358nm bands represent the true absorption peaks for the compound. Thus the initial spectral curve (Curve a) may reflect a considerable degree of oxidation having occurred in the short time it took for the yellow band to be eluted off the chromatogram for the spectrum to be recorded. Once in solution in the cell the stability is apparently increased giving rise to the recorded change after 45 minutes.

(b) Soil Extract Absorption Curves

Having isolated the purified compound as the apparent methyl ester and determined its light absorption behaviour, the information could now be related to the light absorption characteristics of the crude aqueous ring soil extracts.

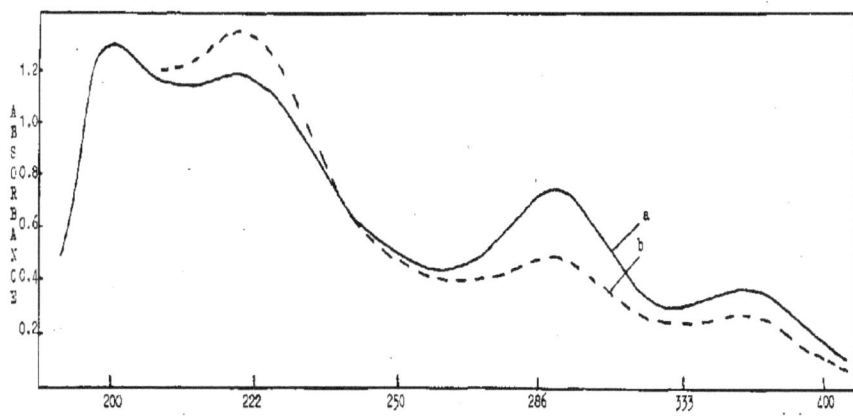

Fig. 9: a) Initial   b) After 45 min in cell

On centrifugation of such mixtures it immediately became evident that material taken from beneath the soil surface gave solutions that were much deeper red in colour than those from surface material. The UV-visible light spectrum of the former typically gave rise to a broad absorption band that extended gradually into the visible region (Fig. 10, Curve a). In contrast, a spectrum of a surface soil extract showed a much more abrupt loss of absorption towards the visible region with the major band occurring in the UV region (Fig. 10, Curve b).

In view of the spectral behaviour of the purified compound, this gross difference may be attributable to the greater degree of oxidation having occurred of the compound towards the soil surface. The overall broadness of the spectra relative to that of the isolated compound is presumed to reflect general decomposition, as opposed to oxidation, incurred while present in the soil. That gradual loss in colour of the extracts towards the ring surface was caused primarily by oxidation was independently inferred from a spectrum recorded on one occasion of an extract of sub-surface material.

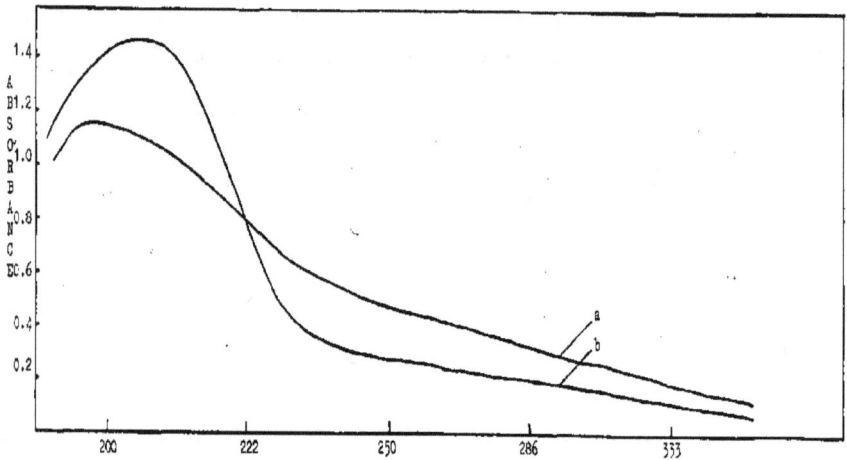

Fig. 10: a) Sub – surface ring soil extract   b) surface ring soil extract

The initial spectrum (Fig. 11, Curve a) showed a broad hump in the region of 286nm attributable to the 290nm absorption band of the pure compound. The compound's instability in its natural aqueous state can be judged from this spectrum since on completion of the recording (taking just two minutes to scan from right to left) it was immediately repeated and this now gave Curve b (Fig. 11). The hump has disappeared and an increase in the region 200-222nm is evident. When more of the same extract was treated with hydrogen peroxide ('40 volume') as an oxidant the red colour was immediately discharged to leave a pale yellow solution whose spectrum (Fig. 11, Curve c) showed the extent to which this decrease/increase in absorption characteristics could be affected. The resulting pale yellow solution on evaporation yielded a whitish solid, and this observation may be of significance when attempting to explain the white surface colour of the ring soil. The strong implication is that the white surface has been caused by a layer of the fully oxidized compound coating the soil. The witnesses' description of a crust-like and crystalline consistency on touching the ring soil may actually be a reference to such a layer.

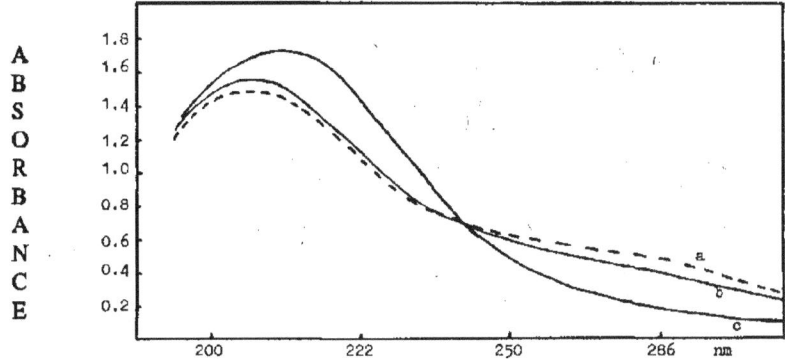

Fig. 11: a) Initial sub-surface extract
b) After 2 minutes in cell
c) After treatment with hydrogen peroxide

It might be instructive here to show a semi-quantitative comparison of surface (*1) and sub-surface (*2) soil absorption curves alongside that of the sample of control soil taken from outside the ring. This is shown in Fig. 12. Each curve was recorded after extraction of 70mg of the relevant soil sample with 0.5ml water. The greater absorption of the ring soil extracts towards the visible region of the spectrum illustrates the degree of impregnation of the soil with the compound.

(c) Chemical Analysis of the Silver Salt

Because of the instability revealed during chromatography and the high degree of decomposition that already occurred to the compound while in the soil, various analytical techniques employed to determine the chemical structure were of limited value. An infra-red absorption spectrum (as KBr disk) of the dried silver salt showed strong hydroxyl (OH) and carboxyl (CO_2^-)

stretching bands at 3,400 and 1600cm^{-1} respectively, with the broad band at 1040cm^{-1} being attributable to C-O stretching (i.e. of C-OH) (Fig. 13).

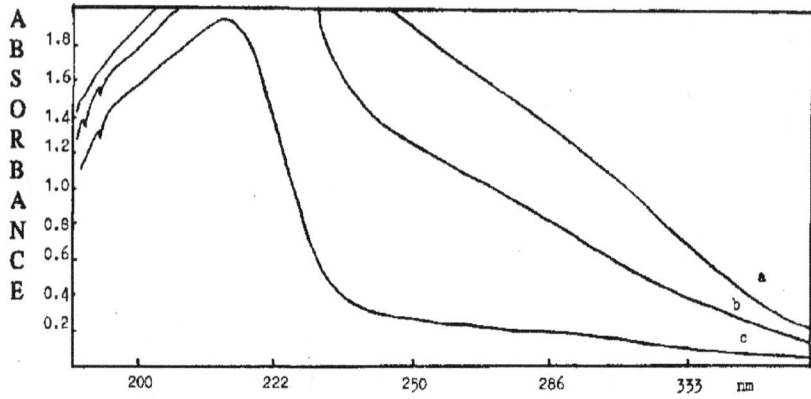

Fig.12: a) Sub-surface ring extract

b) Surface ring extract

c) Control soil extract

A mass spectrum of the crude methyl ester derived from the silver salt was poor owing to its impure nature and possible low volatility and a molecular ion with a characteristic loss of 31 mass units (corresponding to loss of a CH$_3$O$^°$ radical from the ester group) was not apparent in the spectrum. This may also have been due to thermal instability of the ester at the high probe temperature (190°C) used to volatilize the compound. The ion with the highest molecular weight occurred at m/e 395 and this was followed by one of similar but weak intensity at m/e 377. The difference of 18 AMU corresponds to loss of water and would be expected for a compound containing hydroxyl groups. The next ion of any significance occurred at m/e 337 and this also lost water to give a fragment ion at m/e 319. Peaks also occurred at m/e 59 and 57, the former being attributable to the species CO$_2$Me$^+$ that would result if a methyl ester had indeed been formed during treatment with methyl iodide.

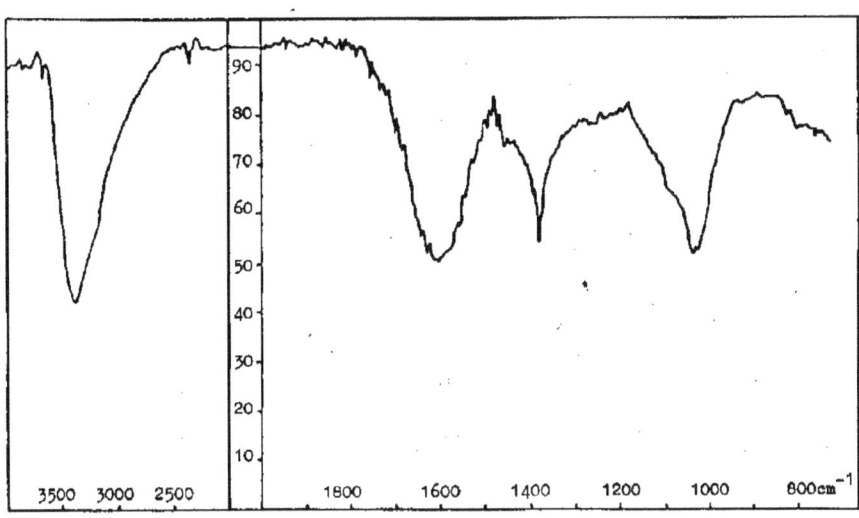

Fig. 13: Infra-red spectrum of silver salt (as KBr disk)

Elemental analysis of the dried silver salt gave 27.55% carbon, 2.40% hydrogen and 3.83% nitrogen. Another performed for ash content gave 27.60% carbon, 2.28% hydrogen, 3.53% nitrogen, and 36.95% ash. Analysis for sulphur and chlorine were both negative. If the residue was assumed to be principally composed of silver oxide (Ag_2O) and the molecular weight of the methyl ester to be in the region 400-500 AMU then a very crude approximation of the empirical formula of the silver salt could be derived as:

$$C_{14}H_{14}N_2O_{12}Ag_2$$

The formula as written has a molecular weight of 618 and is composed of 27.18% carbon, 2.27% hydrogen, 4.53% nitrogen, 31.07% oxygen and 34.95% silver (corresponding to 37.54% silver oxide or ash content). It has to be emphasized that this formula can serve only as a rough guide for the soil compound since, apart from the number of assumptions made in its derivation, it also represents a silver salt containing decomposition and oxidation products as well as of pure compound. The formula's usefulness comes from being able to gauge the relative proportions of each element present in the soil compound. A striking feature is the high oxygen content, which may be accounted for by the presence of two carboxyl groups to pair off with the silver ions, along with a number of hydroxyl groups as indicated by the intensity of the 0-H stretching band in the IR spectrum. The presence of such hydrophilic groups would, in turn, explain the high water solubility exhibited by the soil compound and increase its surfactant properties, underlining further the chemical basis for the hydrophobic nature of the ring soil.

(e) Alleged Local Anaesthetic Effects

The presence of nitrogen in the soil compound may be of significance regarding the alleged local anaesthesia reported. The chemical nature of the nitrogen would have to be determined before this could be properly assessed, however. In order to throw light on this aspect of the case a brief description of the chemical basis of local anaesthesia needs to be presented. The majority of local anaesthetics in use today are tertiary amines that can exist as uncharged or positively charged molecules depending on the pH of the solution and the pKa of the compound. Two examples are Procaine (6) and Lignocaine (7).

H_2N-⬡-$CO_2(CH_2)_2$-NEt_2 (6)

⬡(Me)(Me)-$NHCOCH_2$-NEt_2 (7)

The aqueous solutions of the salts of these compounds with a suitable acid (e.g. hydrochloric acid, HCl) would contain mixtures of the uncharged and charged species, the charge being located at the terminal amino group:

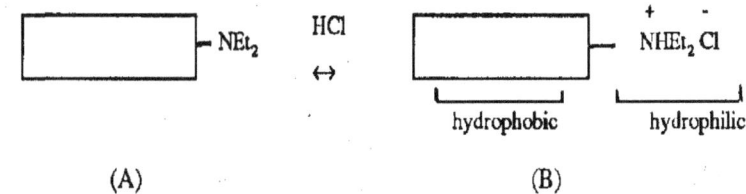

(A)   (B)

Local anaesthetics act at the cell membrane and the block in the conduction of nerve impulses they produce is a result of interference with changes in membrane permeability to potassium and sodium ions. The penetration of a local anaesthetic to its site of action depends upon its ability to cross lipid barriers and the uncharged form (i.e., "A" above) will be able to penetrate lipid layers with ease. Once in the nerve tissues the low pH of the environment generates the charged form (i.e. "B") and it is this species that is responsible for the anaesthetic effect. A mechanism similar to this might apply for the soil compound if the nitrogen present existed as an amine of some kind. Aqueous extracts of the ring soil were found to exude a penetrating putrid odour that on close proximity to hydrochloric acid generated white fumes indicative of a volatile amine present.

The odoriferous component could be extracted out using diethyl ether and the ethereal solution then evaporated down to furnish it in concentrated form. The strong odour of this solution on brief inhalation produced an immediate but short-lived dizzying effect that was rather unpleasant. In an attempt to identify this volatile compound, nitrogen gas was passed through some of the ring soil extract while the gas outlet tube was immersed into a small volume of concentrated hydrochloric acid. Any volatile amines liberated by the soil extract would thus have been trapped in the acid in the form of their hydrochloride salts. After two days the acid solution was removed and evaporated to yield a white crystalline solid which was identified as ammonium chloride from the I.R. spectrum (Fig. 14) (see Fig. 15 for authentic ammonium chloride). This result was odd, since the odour was not recognizably caused by ammonia.

While the nature of the nitrogen remains to be determined, another causative factor for the alleged local anaesthetic effect may simply be the dual hydrophobic-hydrophilic property exhibited by the soil compound and manifested, as described earlier, in the hydrophobicity of the ring soil. In this regard it is of interest to note that a non - nitrogenous compound such as 4-hexyl resorcinol (8) exhibits anaesthetic properties by virtue of this same property and is employed in throat lozenges for this purpose as well as for its antiseptic nature. With regard to the alleged rapidity of the anaesthetic effect we must first assume that the sensation occurred on essentially unbroken skin (e.g. fingertips) on touching the soil. Local anaesthetics are generally administered to mucous membranes or damaged skin and consequently their effectiveness on unbroken skin has been little studied. One such investigation

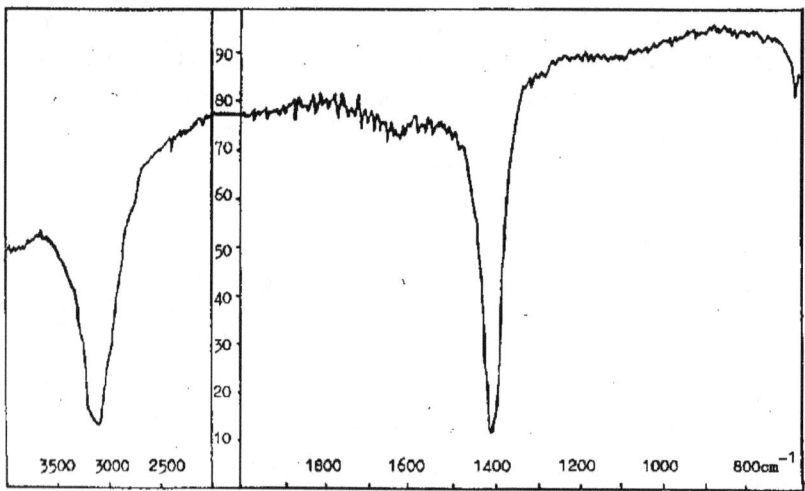

Fig. 14. Infra-red spectrum of volatile amine hydrochloride (as KBr disk).

was carried out in 1957 when a variety of local anaesthetics were administered to the anterior surfaces of the arms and forearms of volunteers and anaesthesia tested with the prick of a needle (Monash 1957). Both the salts of the anaesthetics (charged forms) and the free bases (uncharged forms) were tested, usually as solutions or ointments.

Hydrophilic

$$\text{HO}-\text{C}_6\text{H}_3(\text{OH})-(\text{CH}_2)_4-\text{CH}_3$$

hydrophobic

(8)

Repeatedly swabbing the skin at 15-minute intervals for three hours with 2% solutions of either the salts or the free bases was found to produce no anaesthesia. More success was obtained with an occlusive dressing soaked with 2% solutions of a number of bases. In this case local anaesthesia was produced in 45-60 minutes and was quite prolonged, averaging two to four hours. Perhaps most relevant for this case was the method of continuous contact in which the solution to be tested was placed into a well formed on the skin with petrolatum. A number of anaesthetic salts tested in this way as 2% aqueous solution produced anaesthesia after two hours. These results were interpreted as showing that the free bases (uncharged forms) penetrated the skin more readily than did the salts. The long time interval between application and onset of anaesthesia for both types (45 - 120minutes) showed that the outer layers of the skin acted as an effective barrier to penetration. From this it may be argued that the "immediate" numbing sensation allegedly felt by the witnesses was unlikely. However, the unknown factor is the localized concentration which would appear to be high in the witnesses' case in view of the sensation of crystallinity also noted.

One feature that would be predicted from the alleged anaesthesia is that the ring surface should have been wet or at least moist to provide intimate contact of the substance with the skin. My own attempts to reproduce anaesthesia using an aqueous extract of ring soil on my fingertips left a peculiar sensation which couldn't be described as anaesthetic, although it is possible the solution was not left in contact for long enough.

(f) Chemiluminescence

Although the foregoing is of value in possibly corroborating the alleged local anaesthetic effects, the real significance of the Delphos case lies, I believe, in the instability to air of the compound. The fact that the compound has the rather unusual property of oxidizing to give a fluorescent product leads to the prospect of chemiluminescence (i.e. chemically generated light) being observed. This would, in turn, have implications on the reported "glowing" of the ring soil. To appreciate this fully it will be necessary to describe the mechanism of oxidative chemiluminescence in some detail. Fluorescence is a process in which light of a short wavelength (high energy) is absorbed to promote an electron from the highest bonding orbital to the lowest anti-bonding orbital to produce the first excited singlet state (Fig. 16).

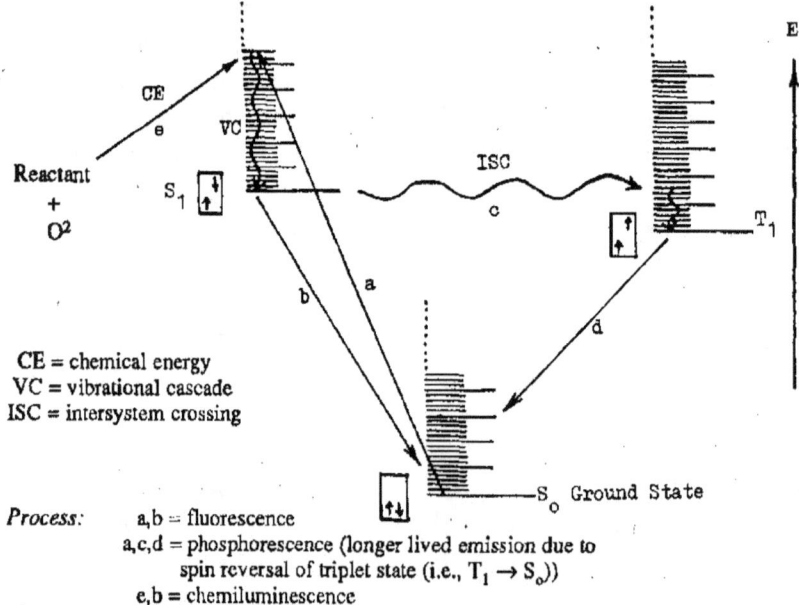

Fig. 16: Fluorescence and chemiluminescence

After 10^{-9} to 10^{-6} seconds the process is reversed and a photon of longer wavelength (in the visible region) is generated. When an organic compound oxidizes to give a product that is fluorescent, emission of light may occur because the energy liberated during oxidation may give rise to a proportion of the product molecules already in their first excited singlet state. Emission would occur when these reverted to their ground state. In such chemiluminescent oxidations the energy supplied should be at least 41 kcal mol^{-1} for emission of red light (700nm) and 65 kcal mol^{-1} for blue light (440nm). Providing that the product molecule is the emitting species (and not some transient chemical intermediate) the fluorescence spectrum of the product should match the chemiluminescence spectrum.

One quantum of light can in principle be emitted by one molecule of reactant and in a perfectly efficient process 1 mole of reactant would generate 1 mole of quanta or 1 einstein. From this it can be calculated that 1kg of perfectly efficient chemiluminescent material might provide 1.47×10^5 lumen hours of light, equivalent to the output of a 40W bulb operating for 13 days.

The quantum yield of a chemiluminescent oxidation is governed by various factors, however, which invariably lead to an overall low efficiency. From the description given above it can be seen that the quantum yield (ϕ_{CL}) for the process will depend on:

i) the fraction of reactant molecules taking the correct chemical pathway in the oxidation (=ϕ_R)

ii) the fraction of product molecules being generated in the first excited singlet state (=ϕ_{ES})

iii) the fluorescence efficiency of the product molecule (if this is the emitting species) (=ϕ_{FL})

from which:

$$\phi_{CL} = \phi_R \times \phi_{ES} \times \phi_{FL}$$

From this relationship it can be seen that all three of these processes must occur significantly for light emission to occur. The product molecule may be highly fluorescent (as is the case for the soil compound) but if the fraction being generated in the excited state during oxidation is small a low chemiluminescent yield will result.

One of the most studied chemiluminescent oxidations is that of the hydrazide, luminol (9). In alkaline dimethyl sulphoxide, luminol reacts with molecular oxygen to produce amino phthalic acid (10) with simultaneous emission of blue light (480nm). The emission spectrum corresponds to the fluorescence spectrum of amino phthalic acid. The chemical yield ((ϕ_R)of the process can be as high as 85%. The fluorescence efficiency (ϕ_{FL}) of amino phthalic acid has been found to be 30%, and its efficiency of formation in the excited state during oxidation of luminol is estimated to be ca. 5%, from which a quantum yield for the process of ca. 1.5% can be calculated. Despite this low figure the emission can be strikingly bright.

(9) → O_2/DMSO → (10) + hν   (11)

One of the most efficient hydrazides yet synthesized is the benzoperylene compound (11) which has a chemiluminescence quantum yield of 7.3%. One can compare this purely chemical process with the enzymically controlled oxidation occurring in the American firefly which is estimated to have a quantum yield of 88%!

The emission wavelength for a chemiluminescent oxidation depends on the energy difference between the first excited singlet state and ground state for the product molecule. In general, the more extensive the chromophore (indicating greater complexity in structure) the smaller is the energy difference giving rise to emission at longer wavelengths. Thus those synthetic chemiluminescent compounds that are structurally relatively simple (e.g. luminol) emit in the blue region of the spectrum.

Another factor that may have an effect on the chemiluminescence spectrum is the chemical environment. Depending upon the solvent, pH, etc., the product molecule may exist in different forms and so cause different emissions. An example is the oxidation of the firefly luciferin analogues (12) in alkaline dimethyl sulphoxide. With low concentrations of alkali, red light (626nm) is produced while higher concentrations cause emission of yellow-green light (562nm). The emitting species for the two processes are the mono-anion (13) and dianion (14) respectively. In addition, the quantum yield for the oxidation varies with the nature of the substituent X.

(12)   (13)   (14)

These examples have all been taken from a number of reviews on chemiluminescence to which the interested reader is referred (Rauhut 1969; White and Roswell 1970; McCapra 1970; Gunderman 1974).

How does all this relate to the ring soil? If a chemiluminescent oxidation had occurred at the soil surface one would expect the soil extracts to contain the fluorescent species responsible for the glow. The surface soil extracts do indeed exhibit a bluish-white fluorescence when viewed under 366nm light. This fluorescence was apparent in all the ring soil samples and was similar to that displayed on the chromatogram of the esterified mixture. A sample of control soil on extraction with water also gave a solution which fluoresced but this was considerably weaker. In an effort to quantify these observations a sample (1.0g) each of the surface ring soil (* 1) and control surface soil were stirred vigorously and at the same rate for 45 min. with deoxygenated water (8.0ml). The mixtures were allowed to settle and the aqueous suspension removed and centrifuged. The control soil gave a very pale yellow solution (for absorption curve see Fig. 12, curve c) while the ring soil solution was deeper yellow in colour (Fig. 12, curve b). The fluorescence spectrum of the undiluted control soil solution showed a maximum at ca. 443nm *(K* (excitation) = 350nm) (Fig. 17). The ring soil solution had to be diluted by a factor of five for the spectrum to be recorded. This showed a similar broad emission but with a maximum at ca. 432nm (Fig. 18). The difference in intensity between the fluorescences of the ring and control surface samples can be seen to best effect if the latter is corrected for the dilution factor and superimposed on the former. The result is shown in Fig. 19.

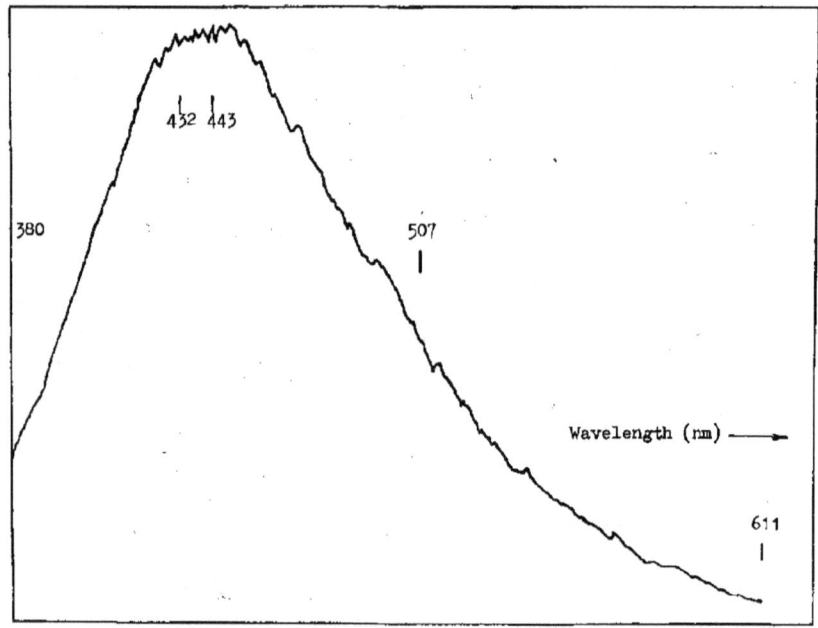

Fig. 17. Fluorescence spectrum of control soil sample.

The ring soil thus fluoresces more brightly than the control soil and this finding is therefore not inconsistent with the claim that the soil was glowing. An attempt was actually made to detect light emission directly from aqueous extracts of the ring soil using a scintillation counter but this was unsuccessful. This may however be attributed to the extensive decomposition and oxidation that has already occurred of the compound, and the fact that these decomposition products gave rise to the broad absorption curves recorded, which might actually hinder any emission from being registered. Coloured impurities are known to weaken or quench the light emission from such reactions. A true demonstration of chemiluminescence may therefore only be possible with purified extracts of the soil compound. The same phenomenon may also explain why the fluorescence of the ring soil extracts appeared suppressed when viewed under 366nm light and is apparently only five times greater than that of unaffected control soil. A much greater difference would have been expected under the circumstances. The fact that the control soil has a similar fluorescence curve to that of the ring soil may at first seem surprising, but I believe it to be purely fortuitous. The reasons for

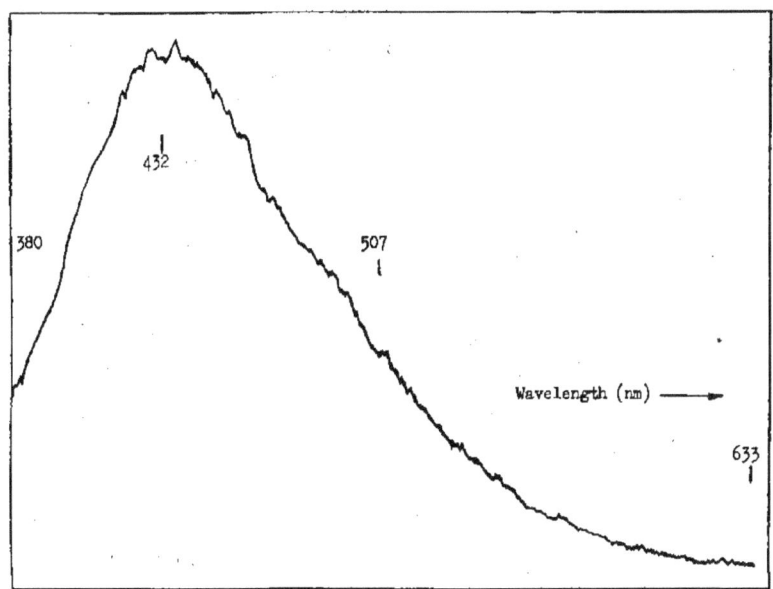

Fig. 18: Fluorescence spectrum of surface ring soil

this are as follows. To begin with the presence of trace organic compounds as appear to be present in the control soil would not be unexpected. That one or more of these might have a chromophoric component leading to a fluorescence in the near UV-visible is likewise not surprising. Secondly, the control soil fluorescence maximum does actually differ from the surface ring maximum while the latter is identical to that of a sub-surface ring sample (*9, Fig. 20). This indicates homogeneity of the ring soil samples. Finally, and most important, a direct association between the ring soil fluorescences and the unstable ring compound which is not present in the control soil can be inferred from another fluorescence spectrum taken of the bands collectively appearing above the purified ester on the chromatogram (i.e. region F, Fig. 8a(ii)). This shows a fluorescence maximum at ca. 429nm (Fig. 21, solvent = spectroscopic acetonitrile). The small difference between this and the aqueous ring soil fluorescences (432-3nm) can be attributed to the different solvent used. Since, on oxidation, the purified ester on the chromatogram gave rise to a fluorescent product indistinguishable from this region and was therefore linked to it through decomposition not affecting fluorescence, this would indicate that the ring soil fluorescences were directly attributable to the unstable compound.

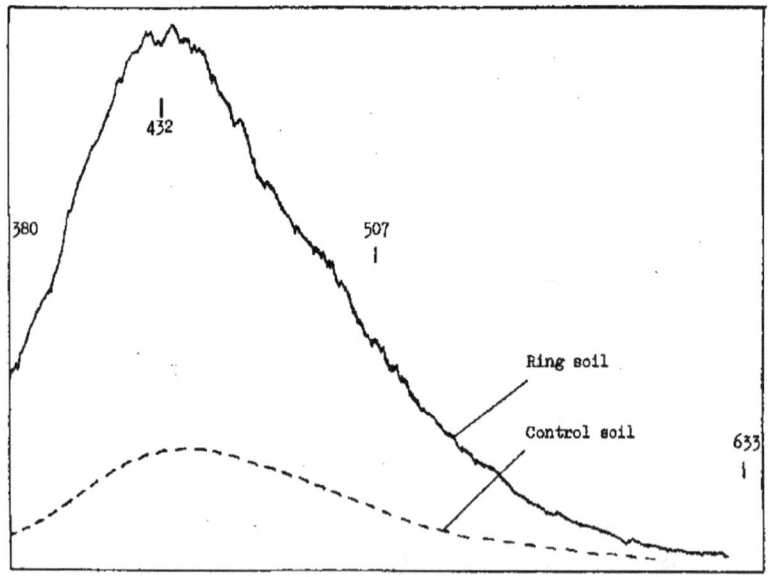

Fig. 19: Fluorescence spectrum of ring soil vs. control soil

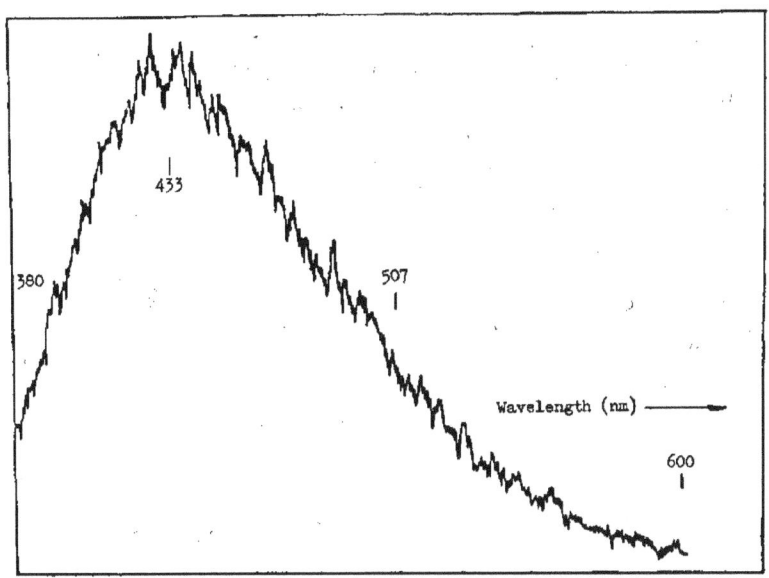

Fig. 20: Fluorescence spectrum of sub-surface ring soil (*9)

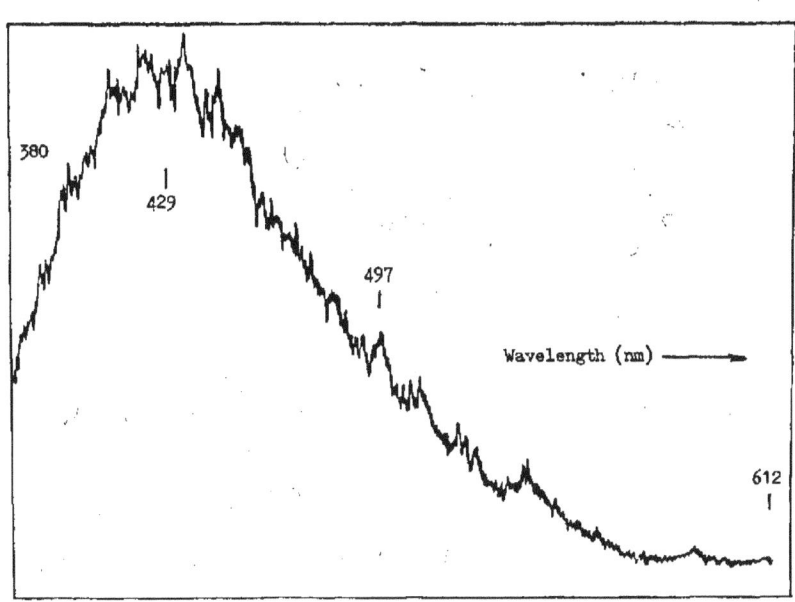

Fig. 21: Fluorescence spectrum of chromatogram region "F"

DISCUSSION

Although the data obtained from the soil analysis is not in any way complete, it does permit the critical examination of certain hypotheses which might be put forward to explain the ring. As I see it, three interpretations are possible.

(a) Hoax

The marked air-sensitivity of the compound and its unusual characteristics render it most unlikely that the ring was the result of a hoax. The ring elongation towards the wind direction on the night of the event would also appear to rule this out. While the possibility of a hoax cannot be completely eliminated until the soil compound has been unambiguously identified, it remains for me, the least plausible of explanations.

(b) Fungal Ring

This explanation appears at first sight to be an attractive one in view of the dimensions and overall appearance of the ring. However, a number of discrepancies become apparent arising from the soil analysis. The chemical nature of the ring soil compound appears to be of a water-soluble alkali metal salt of an organic carboxylic acid. Can such salts be produced by fungi? A common characteristic of fungal rings is that they gradually spread outwards with time having started from a central point of infection. This process would have been expected to continue after discovery of the ring. However, inspection of the ring over a period of six years showed no evidence of growth or even change in shape consistent with such an explanation. Examination of the ring soil did reveal the presence of small clumps of fungal mycelia but the distinct impression was gained that these had resulted due to the enriched organic content of the soil rather than vice-versa. There appeared to be far too much of the compound to have been biologically synthesized by the barely perceptible fungus present. Furthermore, the sample removed by the sheriff the day after the alleged event showed no evidence of fungal growth (Phillips 1979, private communication), suggesting that this arose only after a period of time had elapsed. As in the hoax interpretation a conclusive answer to this question will not be possible until the soil compound has been identified.

(c) *Genuine Sighting*

We come now to the third possible interpretation for the ring, however remote this may at first seem. The information obtained from the soil analysis can, in fact, be assembled together in a manner that leads to a scenario fitting exceedingly well with the sighting report. What follows is my own 'best fit' hypothesis for the ring that takes into account the analysis data and what is generally known about the requirements for chemical light production. It is accepted that this hypothesis is likely to be one of several that may be proposed as an explanation. Nevertheless, I believe it has a degree of symmetry and simplicity that merits presentation, and would welcome any constructive feedback regarding its feasibility. Alternative suggestions or ideas for tests that may be additionally carried out to determine the ring's cause would be equally welcome.

There is present in the ring soil a highly water-soluble organic compound which is potentially chemiluminescent. Although I was unable to demonstrate light emission there are reasons as to why soil material sampled a month after the event might fail such a test. It should also be added that one of the other laboratories involved in investigating the ring soil did apparently succeed in detecting chemiluminescence: "The results initially didn't look very encouraging. On closer examination I did find two curves which did look significant, and which suggest that further study might be fruitful. Most of the light emission appears to be chemiluminescence rather than thermoluminescence. I will try to discover a better way to examine them later" (Phillips 1981: 110). Apparently no further information from this source was available.

The presence of a chemiluminescent compound in the soil in the shape of a ring and the alleged glowing of the object above it suggests that the two may be related. I would propose that the soil compound was responsible for the surface light emission of the object. If this reasoning is followed a number of features of the sighting report now apparently become much clearer. The reported glow between the object and the ground during Ronald's observation would be attributable to the actual deposition of the compound from the object as an aqueous solution. As this met the air beneath, the compound would spontaneously oxidize to create the effect perceived by Ronald. The brightness of the glow suggests that the solution was deposited in a dispersed state such as a spray might be. This would, in turn, lead to two other corroborating features of the ring.

The first is the observed elongation of the ring towards the wind direction on the night in question. The second is the reported blistering of the ring soil by the witnesses on touching the soil. Questioned by Ted Phillips Ronald's father Durel stated, "I never will forget the blisters on the ground, I couldn't describe it, little blisters on it, you know.... little holes, like hail had hit the ground." Similarly with Ronald's mother Erma: "Oh it had a funny feeling.... it felt like it was kind of moist-like.... it was like little blisters like, no holes, bumps, like bumps, blisters..."(Phillips 1981). The moistness of the ring soil reported by Mrs. Johnson would, of course, be entirely consistent with a recent spraying of the soil and would also tie in with the anaesthetic effects felt by both witnesses.

A picture begins to emerge as to what possibly happened that evening. The hovering object of presently unknown origin appears to have contained within its periphery an aqueous solution of an unstable compound whose likely sole function would be light emission. Since oxygen is a requirement a mechanism for controlling the emission might come from regulating the absorption of air through its outer surface. The latter would have to be constructed from a transparent material permeable to oxygen (Fig. 22). Some of the solution was deposited into the ground while the object positioned itself under a tree (to possibly avoid observation from the air?). Both the surface emission of the object and that between it and the ground occurred simultaneously according to Ronald. This may suggest that oxygen absorption through the surface occurred in response to the solution being ejected from underneath. The rumbling noise heard at the same time "like a washing machine that vibrates" might be associated with the manner in which the deposition occurred.

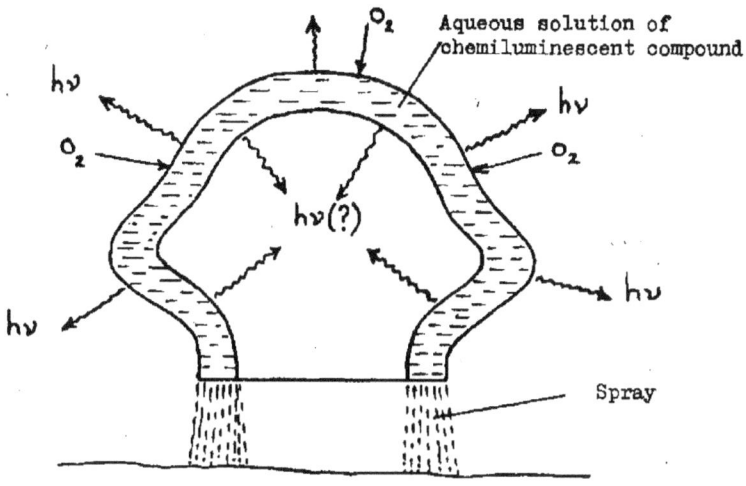

Fig. 22: Possible chemical emission from the UFO

Once enough of this solution was deposited the object departed after which the Johnson family approached the ring area. The freshly laid material was still oxidizing and emitting light at the ring surface which was observed and described as "glowing" brightly enough to prompt Mrs. Johnson to quickly fetch her camera and take the photograph (Fig 4). Eventually the oxidation would have gone to completion leading to cessation of the glow and the production of the reflective white coating that is visible in the photograph taken the day after by the sheriff. Since the latter is likely to have been visible on subsequent moon lit nights, it is this author's opinion that the claim by the witnesses that the ring glow was still visible 24 hours after the event may actually be erroneous. As regards the colour of the alleged UFO, from the description given earlier on oxidative chemiluminescence the emission spectrum can be equated to the fluorescence spectrum of the oxidation product which in this case is a blue emission (432nm) in aqueous solution. Ronald initially described the UFO as multicoloured with blue, red, and orange glows about its surface. Two possibilities come to mind for explaining this. The first is that the precise chemical environment within the object might have produced a different form of the product molecule resulting in emission of the red-orange light. An alternative, and more satisfying interpretation, is that while hovering there was a non-uniform absorption of oxygen about the object's surface resulting in patchy emission of the blue light. This would have had the effect of illuminating non-emitting regions whose observed colour would be concentration dependent and arise from the tail absorption of the 358nm band of the pure compound. That such colours might result was suggested from concentrating down the crude yellow acetone solution obtained from the silver salt esterification process. This invariably resulted in red gums being obtained. After departing the object was described by the family as receding into the distance bearing "the colour of an arc welder." One supposes here that the arc-welding process referred to is the typical mild-steel one that can be found in operation in any garage workshop. Such welding processes commonly emit intense bluish-white light, and having seen one in operation myself the similarity between the emission and the ring soil fluorescence was very striking.

FURTHER INVESTIGATION

The foregoing represents an attempt to investigate and explain an exceptional UFO trace case report. The original investigator of this case Ted Phillips has catalogued over 4000 such cases that vary in their significance and evidential content, but none can match this particular one in this author's judgment. A preliminary version of this report was published in the relatively low readership Journal of UFO Studies in 1989, following which a U.S. analyst Phyllis Budinger received additional Delphos ring soil material to carry out an independent study in 1999. She subjected the organic material to a variety of machine-based HPLC, GC, IR and NMR analyses to acquire an overall picture of what the ring soil contained. Her results were published the same year and an on-line version of her report is linked here:

http://documents.theblackvault.com/documents/Budinger/UT001.pdf

She confirmed that a deposition from an external source was most likely to have occurred, and correctly identified the soil compound as a highly water-soluble organic carboxylic acid derivative. She was further able to reproduce the silver salt precipitation process that I had used to isolate the silver salt of the compound. However, because she was not versed in synthetic chemistry she did not go further and perform the vital chemical conversion of the silver salt into the methyl ester in order to isolate and characterize the main, potentially chemiluminescent component present in the mixture. Because the soil material had been sampled by Phillips a month after the event and then kept in storage for many years, it is very likely that natural breakdown of the compound to some degree would have occurred whilst in the soil, in other words it would have undergone a humification process. It should be emphasized that any organic substance placed within soil will eventually suffer such a breakdown with time, being caused by the presence of indigenous micro-organisms, as well as through natural oxidative and hydrolytic processes. Since all she could mainly identify from her analyses were chemical groups that correlated with this humification she characterized the ring soil compound as being a deposited 'humate' substance. Moreover, since it was water soluble, she narrowed down her assessment by referring to the soil compound as a 'fulvic acid'. The latter is the name normally used to describe a naturally occurring water-soluble humic acid. Unfortunately, the assignment of this name subjectively implies the ring soil compound to therefore be natural in origin, which assessment this author strongly disagrees with.

One other compound Budinger determined to be present in the ring soil was a high concentration (5%) of oxalic acid (15). This is a toxic compound that I had missed during my analysis. Its presence can again be viewed as unusual, and certainly more so if the ring was in any way to be proposed as being 'natural' in origin. However, in this author's opinion it may potentially represent another vital piece of evidence that actually supports the witnesses' testimony of a luminescent object. This is because oxalic acid is a well known key component of highly efficient chemiluminescent reactions! For this purpose oxalic acid can readily be converted into suitable diesters which on treatment with alkaline hydrogen peroxide in the presence of an appropriate fluorescent compound can lead to light emission with quantum yields of up to 25%. This is the basis of the commercially available Cyalume Glow Stick (or Light Stick) originally developed by the American Cyanamid company. More information on this product can be found at the following link:

http://en.wikipedia.org/wiki/Glow_stick.

It must be emphasized that the mechanism of this chemiluminescent process is very different from the one postulated for the purified soil compound, since it requires hydrogen peroxide as the oxidant, rather than aerial oxygen. Nevertheless it is not inconceivable that the use of oxalyl derivatives might be incorporated into chemiluminescent reactions that *were* able to use molecular oxygen as the oxidant. In this regard it is interesting to note that the commercial synthesis of hydrogen peroxide involves bubbling molecular oxygen into a solution of 2-ethyl-9,10-dihydroxyanthracene (16) to oxidize it into the corresponding 2-ethylanthraquinone (17) while generating free hydrogen peroxide as a side product which is then extracted and purified. By postulating a suitable oxalyl diester version of the substrate (16) contained within the UFO the absorption of aerial oxygen would enable *in-situ* generation of the required hydrogen peroxide and thereby potentially result in a highly efficient cyalume emission process strong enough to have caused Ronald's temporary blindness. It follows that the deposition of any un-reacted solution of this postulated oxalyl diester into the ring soil would lead to the formation of free oxalic acid that was eventually identified by Budinger through soil breakdown.

Following the Budinger analysis, a more complete report on the Delphos encounter incorporating all the available laboratory data was published by the UFO Research Coalition (Phillips, 2002). This was edited by Jennie Zeidman and included a Scientific Results Summary by Professor Michael Swords of Western Michigan University in its concluding section. The consensus of opinion was - and remains - that the case report cannot adequately be explained in prosaic terms. In view of the obvious ramifications should the case implicate a genuine deposition by an unknown aircraft it is this author's opinion that the information presented here should be submitted for wider scrutiny and appraisal by the scientific community at large.

$$HO-C(=O)-C(=O)-OH$$

(15)

(16) 9,10-dihydroxy-2-ethylanthracene $\xrightarrow{O_2}$ (17) 2-ethylanthraquinone + H_2O_2

(16)   **(17)**

ACKNOWLEDGEMENTS

I wish to thank Dr. Mark Rodeghier and Jerome Clark of the Center for UFO Studies, Chicago, Illinois, 60631 for their helpful advice in the preparation and checking of this manuscript.

REFERENCES

Gunderman, K.D.

    1974   Recent Advances in Research on the Chemiluminescence of Organic Compounds. In *Topics in Current Chemistry 46: Photochemistry* Berlin: Springer-Verlag. pp. 61-139.

McCapra, F.

    1970   The Chemiluminescence of Organic Compounds. *Pure and Applied Chemistry,* Vol. 24: 611.

Monash, S.

    1957   Topical Anaesthesia of the Unbroken Skin. *Archives of Dermatology.* Vol.76: 752.

Phillips, Ted

    1972a   Landing Report from Delphos. *Flying Saucer Review Case Histories,* No. 9: 4-10.

1972b	Landing Report from Delphos. In 1972 *MUFON UFO Conference Proceedings.* Quincy, 111.: Midwest UFO Network, pp. 53-60.
1981	Close Encounters of the Second Kind: Physical Traces - A Case in Point, Delphos, Kansas. In *1981 MUFON UFO Symposium Proceedings.* Seguin, Tex.: Mutual UFO Network, pp. 105-129.
2002	Delphos – A Close Encounter of the Second Kind: UFO Research Coalition, Fairfax, VA 22032.

Rauhut, M.M.

1969	Chemiluminescence from Concerted Peroxide Decomposition Reactions, *Accounts of Chemical Research,* Vol. 2: 80.

Sturrock, P.A.

1997	Physical Evidence Related to UFO Reports: The Proceedings of a Workshop Held at the Pocantico Conference Center, Tarrytown, New York, September 29 - October 4, 1997. In *Journal of Scientific Exploration,* Vol.12, No.2, pp. 179 - 229, 1998.
2000	The UFO Enigma: A New Review of the Physical Evidence: Warner Books, 1271 Avenue of the Americas, NY 10020.

White, E.H., and D.F. Roswell

1970	Chemiluminescence of Organic Hydrazides. *Accounts of Chemical Research,* Vol. 3: 54.

ABOUT THE AUTHOR

The author is a British born scientist from Turkish parentage. After receiving his B.Sc (Hons) in chemistry from Queen Mary College, London University, he stayed on to earn a Ph.D in the organic synthesis of unstable carotenoid pigments before moving to Oxford and Nottingham Universities to carry out postdoctoral research in other areas of chemistry. He subsequently found long term employment as a pharmaceutical development chemist which led to him being named inventor and co-inventor of numerous patents awarded for commercially important processes. The author's early interest in astronomy indirectly led to his curiosity into the UFO phenomenon, and once he became adept in chemistry he was keen to use his expertise to uncover knowledge on the phenomenon if at all possible. This book is the result of that endeavour.

www.ingramcontent.com/pod-product-compliance
Lightning Source LLC
Chambersburg PA
CBHW051716170526
45167CB00002B/679